Schriftenreihe aus dem Institut für Strömungsmechanik

Herausgeber
J. Fröhlich, R. Mailach, S. Odenbach

Institut für Strömungsmechanik
Technische Universität Dresden
D-01062 Dresden

Band 20

Tian Ma

A contribution to turbulence modelling in bubbly flows

TUD_press_
2017

Die vorliegende Arbeit wurde am 19. September 2016 an der Fakultät Maschinenwesen der Technischen Universität Dresden als Dissertation eingereicht und am 08. Mai 2017 erfolgreich verteidigt.

This work was submitted as a PhD thesis to the Faculty of Mechanical Science and Engineering of TU Dresden on 19th September 2016 and successfully defended on 08th May 2017.

Gutachter | Reviewers
1. Prof. Dr.-Ing. habil. Jochen Fröhlich
2. Prof. Dr.-Ing. habil. Suad Jakirlić

Bibliografische Information der Deutschen Nationalbibliothek
Die Deutsche Nationalbibliothek verzeichnet diese Publikation in der Deutschen Nationalbibliografie; detaillierte bibliografische Daten sind im Internet über http://dnb.d-nb.de abrufbar.

Bibliographic information published by the Deutsche Nationalbibliothek
The Deutsche Nationalbibliothek lists this publication in the Deutsche Nationalbibliografie; detailed bibliographic data are available in the Internet at http://dnb.d-nb.de.

ISBN 978-3-95908-115-3

bei Thelem Universitätsverlag
und Buchhandlung GmbH & Co. KG
Bergstr. 70 | D-01069 Dresden
http://www.tudpress.de

Gesetzt vom Autor. | Typeset by the author.
Printed in Germany.

Technische Universität Dresden
Fakultät Maschinenwesen

A contribution to turbulence modelling in bubbly flows

Dissertation

zur Erlangung des akademischen Grades

Doktor-Ingenieur (Dr.-Ing.)

von

Dipl.-Ing. Tian Ma

Gutachter:	1.	Prof. Dr.-Ing. habil. Jochen Fröhlich
	2.	Prof. Dr.-Ing. habil. Suad Jakirlić
Tag der Einreichung:		19.09.2016
Tag der Verteidigung:		08.05.2017

"临阵磨枪，不快也亮。"

"Brighten the spear before fighting; even not sharp, it is brilliant."

Jun Zhu, my middle school teacher

This thesis is dedicated to my loving Dad and Mom.

Acknowledgements

First of all, I would like to express many thanks to Prof. Jochen Fröhlich for accepting me as a PhD student in 2013. I have learned many things from him in the last years, not only the detailed scientific points in the present thesis, but also how to work as a scientific associate. For me, the regular meetings in his office have always been a great source of motivation. I really enjoyed it.

I am also very grateful to Prof. Suad Jakirlić, who I know from a nice conference in Marbella, for accepting the part as the second examiner of my thesis.

Dr. Dirk Lucas is my scientific supervisor in Helmholtz-Zentrum Dresden-Rossendorf where I am working. Without his support in many ways, this work would have not been possible. He taught me a lot through numerous discussions and offered a great deal of physical insight that gave me a lot of inspiration for some of the key points during my work.

I thank Dr. Thomas Ziegenhein and Dr. Yixiang Liao. It is my great pleasure to work with them and share some free time with them in Dresden. I extend my thanks to the many colleagues and students from my group, especially Dr. Eckhard Krepper, Dr. Friederike Gauß, Susann Hänsch, and Pavel Apanasevich, for their fruitful discussions and support. Prof. Akio Tomiyama, Prof. Niels Deen, Dr. Claudio Santarelli, and Dr. Ivo Roghair are acknowledged for providing their data in electronic form.

Finally, above all, I wish to thank my Dad and Mom for their support, patience, and love over the years.

Abstract

This thesis presents the investigation of turbulence modelling in dispersed bubbly flows. Euler-Euler simulations are performed in bubble columns and bubbly channel flows using different turbulence models. The focus is first set on the large eddy simulation performed for a bubble column with gas homogeneously distributed and a bubble plume. Instantaneous and statistical data obtained from the simulations are presented and compared with experimental data. The impact of the modelling on the liquid velocity fluctuations is investigated and the energy spectrum obtained from the resolved velocity is discussed.

To cover a wide range of reliable performance with scale resolving simulation, scale-adaptive simulations are conducted for a square cross-sectioned bubble column. The main objective is to investigate the potential of this approach for the prediction of bubbly flows with anisotropic liquid velocity fluctuations. Experimental data and Euler-Euler large eddy simulation are used for comparison.

In addition to scale resolving simulations, the thesis addresses a complete framework to develop a bubble-induced turbulence model embedded in Reynolds-averaged Navier-Stokes model. Budgets of the turbulent kinetic energy from direct numerical simulations of dispersed bubbly channel flows are used in the model development. Appropriate length and time scales for the bubble-induced turbulence are determined and the corresponding source terms in the Shear Stress Transport Model are derived from the data. This yields a closure which can be readily used in many existing methods. The model is validated by dedicated Euler-Euler simulations and expected to provide reliable performance in a wide range of applications.

Zusammenfassung

Diese Arbeit befasst sich mit der Turbulenzmodellierung disperser Blasenströmungen. Unter Verwendung unterschiedlicher Turbulenzmodelle werden Blasensäulen und Blasen-Kanalströmungen mit der Euler-Euler-Formulierung simuliert. Der Fokus liegt zunächst auf der Large Eddy Simulation einer Blasensäule mit einer homogenen Gasverteilung und einer Blasenfahne. Zeitabhängige und statistische Ergebnisse der Berechnungen werden analysiert und mit experimentellen Daten verglichen. Die Auswirkungen der Modellierung von Geschwindigkeits-Fluktuationen der Flüssigphase werden untersucht und das Energiespektrum der Geschwindigkeitsverteilung wird diskutiert.
Für eine Blasensäule mit quadratischem Querschnitt werden neben der Large Eddy Simulation Scale Adaptive Simulations durchgeführt, um einen weiteren Bereich der Scale Resolving Simulations abzudecken. Hierbei ist das zentrale Ziel, das Potential dieses Ansatzes für die Verhaltens-Vorhersage von Blasenströmungen mit anisotropen Geschwindigkeitsfluktuationen der Flüssigphase zu untersuchen. Zum Vergleich werden experimentelle Daten und eine Euler-Euler Large Eddy Simulation herangezogen.
Neben der skalenaufgelösten Simulation, wird in dieser Arbeit die Grundlage für ein Modell zur Beschreibung der Blasen-induzierten Turbulenz ausgearbeitet, das in die Reynolds-gemittelten Navier-Stokes-Gleichungen integriert wird. Zur Entwicklung dieses Modells werden Ergebnisse aus der direkten numerischen Simulation einer dispersen Blasen-Kanalströmung verwendet. Für die Blasen-induzierte Turbulenz werden geeignete Längen und Zeitskalen berechnet. Der entsprechende Quellterm im Shear Stress Transport Modell basiert auf den Daten der direkten numerischen Simulation. Dies ergibt eine Lösung des Schließungsproblems, die in vielen existierenden Modellen leicht angewendet werden kann. Die Validierung des Modells findet mit entsprechenden Euler-Euler-Simulationen statt. Es wird erwartet, dass das neue Modell in einem großen Anwendungsbereich eine zuverlässige Leistung bietet.

Nomenclature

Latin symbols

A	surface area
Ar	Archimedes number
C_L	lift force coefficient
C_W	wall force coefficient
C_D	drag coefficient
C	model coefficient in the dynamic Smagorinsky model
C_s	model coefficient in the Smagorinsky model
C_{TD}	turbulent dispersion force coefficient
C_I	coefficient in the modelled k-source for BIT
C_ε	coefficient in the modelled ε-source for BIT
C_ω	cross-diffusion in ω-equation
C_μ	coefficient in eddy-viscosity formula
$C_{\varepsilon 1}$, $C_{\varepsilon 2}$	coefficients in the modelled dissipation equation
D_k	diffusion in k-equation
D_ω	diffusion in ω-equation
d_p	bubble diameter
Eo	Eötvös number
$\boldsymbol{F}$	force
$\boldsymbol{F}_D$	drag
$\boldsymbol{F}_L$	lift force
$\boldsymbol{F}_W$	wall force
$\boldsymbol{F}_{TD}$	turbulent dispersion force
g	gravitational acceleration
k	turbulent kinetic energy
M	interfacial momentum forces
Mo	Morton number
$\boldsymbol{n}$	normal boundary vector
N_p	number of bubbles
p	pressure
Re_p	Bubble Reynolds number
Re_b	bulk Reynolds number
Re_τ	bubble Reynolds number
$\boldsymbol{S}$	strain rate tensor
S_k	BIT source in k-equation
S_ω	BIT source in ω-equation

t	time
$\boldsymbol{u}$	velocity vector with Cartesian components
$\boldsymbol{u}'$	velocity fluctuation vector with Cartesian components
V	volume
$\boldsymbol{x}$	coordinate vector with components x, y, z

Greek symbols

α	void fraction
Δ	grid size
μ	dynamic viscosity
ν	kinematic viscosity
ν_t	turbulent (eddy) viscosity
ν_{sgs}	subgrid turbulent viscosity
ν_{eff}	effective viscosity
ν_{mol}	molecular viscosity
ν_{bub}	bubble-induced viscosity
ρ	density
∇	Nabla operator
σ	surface tension
Π_k	production in *k*-equation
Π_ω	production in ω-equation
ε	turbulent eddy dissipation rate
ε_k	dissipation in *k*-equation
ε_ω	dissipation in ω-equation
$\boldsymbol{\tau}_t$	unresolved stress tensor
ω	turbulent eddy frequency
Φ	liquid indicator function

Subscripts

b	bulk flow
G	gas
K	phases (*L*, liquid; *G*, gas)
L	liquid
p	bubble/particle
r	relative velocity related quantity
ref	reference quantity
t	turbulence related quantity

Abbreviations

BIT	bubble-induced turbulence
CFL	Courant-Friedrichs-Lewy number
DNS	direct numerical simulation
FFT	fast Fourier transformation
LDV	laser Doppler velocimetry
LES	large eddy simulation
PIV	particle image velocimetry
RANS	Reynolds-averaged Navier-Stokes
SAS	scale adaptive simulation
SGS	subgrid scale
SIT	shear-induced turbulence
SRS	scale resolving simulation
TKE	turbulent kinetic energy

1 Introduction

1.1 Background and motivation for the thesis

Many technical flows feature a continuous liquid phase and a dispersed gaseous phase. The turbulence in the liquid phase is an important issue in such multiphase flows, as it has a strong influence on the local distribution of the dispersed phase and on the bubble size by bubble fragmentation and coalescence (Liao, et al., 2015). The distribution of the dispersed phase on the other hand impacts on the large-scale velocity field and the small-scale turbulence, so that a very complex interaction mechanism arises (Balachandar & Eaton, 2010). Compared to the liquid phase, the turbulence in the dispersed gas phase is of little relevance and is commonly neglected (Ishii & Hibiki, 2011). This is due to the low density of the gas and the small dimensions of bubbles.

The bubbles interact with the liquid phase and strongly modify the statistical and structural characteristics of the flow, even if the gas void fraction is as low as 0.1% (Prakash, et al., 2016). If the background flow is turbulent, the flow situation is then more complicated in such cases as illustrated by Hosokawa & Tomiyama (2010): on one hand, with the introduction of bubbles the turbulence is attenuated due to disintegration of the eddies and disruption of the energy transfer; on the other hand, liquid turbulence is augmented due to the bubble wakes; furthermore, the mean flow profile of liquid could also be modified by bubbles, which changes the level of shear-induced turbulence (SIT). The combination of various effects can yield an overall enhancement or attenuation of the turbulence intensity in the liquid carrier phase (Gore & Crowe, (1989), Tanaka & Eaton (2008)).

Bubbly flows have been extensively investigated using experimental techniques (Serizawa et al. (1975); Lance & Bataille (1991); Hosokawa & Tomiyama (2004); Ziegenhein & Lucas (2016); Barbosa et al. (2016)). In particular, the gas void fraction, liquid mean flow, and various turbulence parameters were measured and analysed. It was observed that modification of carrier flow by bubbles depends on the flow conditions (e.g. flow Reynolds number) and on the characteristics of the bubbles (e.g. bubble size and total gas hold-up).

Bubbly flows can also be investigated numerically by various approaches at different levels of detail. These approaches can be classified as direct numerical simulations (DNS), Euler-Lagrange (EL), and Euler-Euler (EE) methods. DNS use interface resolving method to treat the phase boundary and fully resolve the bubble geometry, as well as the interfacial effects. In contrast to resolving the interface, an Euler-Lagrange method considers the bubbles as point force and tracks them in a Lagrangian manner, while liquid phase is solved by phase-averaged equations in an Eulerian manner.

For large scale simulations, the Euler-Euler approach is the only viable framework, where separate balance equations with additional modelling terms are solved for the different phases. The current state-of-the-art of such approaches is discussed in a number of books e.g. by Drew & Passman (1998), Prosperetti & Tryggvason (2007), and Ishii & Hibiki (2011). Particularly, for turbulence modelling within this framework, many attempts were made in the past. For the continuous liquid phase, the first works were based on a linear superposition of the contributions to the turbulent viscosity due to shear- and bubble-induced turbulence (BIT) given by Sato & Sekoguchi (1975) and Drew & Lahey (1981). In such models the bubbles influence on liquid turbulence is included by an additional extra term contributing to the bubble-induced viscosity and the SIT is determined using an unmodified Reynolds-averaged Navier-Stokes (RANS) or scale-resolving simulation (SRS) models. Notwithstanding certain success in the prediction of phase and velocity distributions using EE-RANS in pipe flows (Krepper, et al., 2005) and in bubble columns (Tabib, et al., 2008), application of such algebraic approaches was quite limited due to the too simplistic linear superposition of viscosity from BIT and SIT. In the present study, the performance of this approach will be systematically checked in the framework of EE-SRS to get a complete overview.

Modelling of turbulence modulation in bubbly flows can also be identified from the rigorous derivation of the basic equations of turbulence in gas-liquid flow, which was proposed by Kataoka & Serizawa (1989) for bubbly flow. The formulation they derived is based on the ensemble averaging of local instantaneous conservation equations. It employs a single-phase flow representation but includes the influence of the bubbles by means of additional interfacial source terms in the balance equations both for k, the turbulent kinetic energy (TKE) and ε, the dissipation rate of TKE. The complete budgets of TKE and dissipation rate of TKE allow a statistical quantification of the mechanisms involved, hence yielding information about their relative importance in a particular flow considered. Since then, their formulation has been usually employed in the framework of RANS/SRS modelling for bubbly flow, various forms of the BIT models have been proposed, e.g. by Troshko & Hassan (2001), Politano et al. (2003), de Bertodano et al. (1994), and Morel (1997). However, due to the lack of reliable measurements of exact interfacial terms derived by Kataoka & Serizawa (1989), the forms of such models were mostly based on *ad hoc* adjustments of the model constants involved in the source terms and different choices of the time scale employed in the BIT term of the dissipation equation to reproduce the flow fields. These issues motivate the present work to do a detailed investigation to figure out such objectives and to improve the modelling of BIT.

During the last two decades, direct numerical simulations (DNS) of bubbly flows have progressed rapidly (Ervin & Tryggvason (1997); Bunner & Tryggvason (2003); Roghair et al. (2011a); Bolotnov et al. (2011); Dabiri et al. (2013); Lu & Tryggvason (2013)). The arrival of DNS has shed some light on the accurate evaluation of TKE budgets derived from DNS data and this provides a route toward the modelling of BIT source terms and the assessment of the available models. However, attempts to use DNS data to help modelling of BIT are still rare (Ma, et al., 2015). The DNS data of bubbly flow are usually solely treated as validation database for mean flow and turbulent parameters, but not as a direct contribution for the improvement of an explicit term in the equations. The most intense effort until 2015, if not the only, with using DNS data for the evaluation of the budget terms appears to have been by Ilić (2006) and Erdogan & Wörner (2014). They tested some BIT models from literature using the achieved DNS data for the same configurations of their DNS and found none of the BIT models they considered performed very accurately for all the cases.

However, to use the DNS data of bubbly flow to improve BIT model was not done by Ilić (2006) and Erdogan & Wörner (2014). The strategy for this issue can be followed by the long-time experience in RANS modelling for single-phase flow, where DNS data are used as a basis for development of improved RANS models and have resulted in significant progress (see e.g. Mansour et al. (1988), Yang & Shih (1993), Rodi & Mansour (1993), Jakirlić &

Hanjalić (2002), and Jakirlić & Jovanović (2010)). Among these, the pioneering works of Mansour et al. (1988) and Rodi & Mansour (1993) deserve special attention because they are quite close in spirit to the present study, although both works are related to single-phase flow. Mansour et al. (1988) computed both the k-budget and ε-budget using the DNS data of Kim et al. (1987) for a channel flow. Then they used an *ad hoc* eddy viscosity damping function evaluated from the given DNS data in the computation and found an improvement in the prediction of the k-profile compared to other existing near-wall turbulence models. In the latter work given by Rodi & Mansour (1993), term-by-term modelling of the ε-equation was attempted on the basis of DNS data and a new wall damping function using curve-fitting to the high-Reynolds-number channel flow data was determined.

The works of single-phase flow mentioned above give the motivation and guidance of the present work, which is attempted to use the DNS data of complete two-fluid system to model the interfacial terms responsible for BIT in the framework of EE-RANS.

1.2 Outline of the thesis

The remainder of this thesis is organized in five chapters. In Chapter 2, the fundamentals of the modelling approaches used for bubbly flows are introduced. It contains two parts in the framework of Euler-Euler approach: the interfacial force modelling and the turbulence modelling. Chapter 3 presents and discusses different aspects of the results arising from the large eddy simulations (LES) conducted for various bubble columns. This chapter combined with Chapter 4, which demonstrates the instantaneous and statistical results of a scale adaptive simulation of a bubble column, and provides the state-of-the-art of the EE-SRS. Details of benefits and limitations about this approach are discussed. The results reported in these two chapters are based on Ma et al. (2015a), Ma et al. (2015b), and Ma et al. (2016). Besides SRS, the RANS approach is also investigated in the same work. In Chapter 5, budgets of the turbulent kinetic energy from direct numerical simulations of disperse bubbly channel flows are used to develop a new model for bubble-induced turbulence in the EE-RANS framework. The model is validated by dedicated simulations and ready for use. Part of the results in Chapter 5 have been presented in Ma et al. (2017). Conclusions of the thesis and future work are finally presented by Chapter 6.

2 Physical modelling for bubbly flows

2.1 Euler-Euler approach

In this work the Euler-Euler two-fluid model is used. The conservation equations are discussed in detail in a number of books, such as Ishii & Hibiki (2011); Drew & Passman (1998), and a broad consensus on this model has been reached. The governing equations of this approach are the continuity and momentum equations for each phase employed here, without mass transfer between the phases

$$\frac{\partial}{\partial t}(\alpha^K \rho^K) + \nabla \cdot (\alpha^K \rho^K \overline{\boldsymbol{u}}^K) = 0, \tag{2.1}$$

$$\begin{aligned}\frac{\partial(\alpha^K \rho^K \overline{\boldsymbol{u}}^K)}{\partial t} &+ \nabla \cdot (\alpha^K \rho^K \overline{\boldsymbol{u}}^K \overline{\boldsymbol{u}}^K) \\ &= -\nabla \cdot (\alpha^K \mu^K \overline{\boldsymbol{S}}^K) - \alpha^K \nabla \overline{p} + \alpha^K \rho^K \boldsymbol{g} + \boldsymbol{M}^K - \nabla \cdot (\alpha^K \boldsymbol{\tau}_t^K).\end{aligned} \tag{2.2}$$

Here, the superscript K denotes the different phases, with α, ρ, μ and $\overline{\boldsymbol{u}}$ being the volume fraction, density, molecular viscosity, and some resolved velocity (needs to be specified in its meaning), respectively, and $\overline{\boldsymbol{S}}$ is the strain rate tensor. The superscript K can be L to designate the liquid and G to designate the gas. The vector $\boldsymbol{M}$ represents the sum of all interfacial forces acting between the phases. Additionally, all phases share the same pressure field, namely, $\overline{p}^L = \overline{p}^G = \overline{p}$. The unresolved stress tensor $\boldsymbol{\tau}_t$, and all interfacial forces have to be modelled. The applied modelling is discussed below. Part of this chapter is based on Ma at el. (2015a).

2.2 Modelling interfacial forces

In the Euler-Euler approach, the momentum exchange between gas and liquid is modelled through exchange terms in the separate momentum equations of the different phases. Theses interfacial forces $\boldsymbol{M}$ (Figure 2.1) consist of different contributions, namely drag force $\boldsymbol{F}_D$, lift force $\boldsymbol{F}_L$, wall force $\boldsymbol{F}_W$, and turbulent dispersion $\boldsymbol{F}_{TD}$, and is given by

$$\boldsymbol{M}^G = -\boldsymbol{M}^L = \boldsymbol{F}_D + \boldsymbol{F}_L + \boldsymbol{F}_W + \boldsymbol{F}_{TD}\,. \tag{2.3}$$

Details of such models including a complete description of all interfacial transfer were published by Ishii & Hibiki (2011). There is still no agreement in the community on the closures to be used at best. The corresponding interfacial transfer models employed here are briefly recalled. A more detailed discussion on all these models can be found in the original papers.

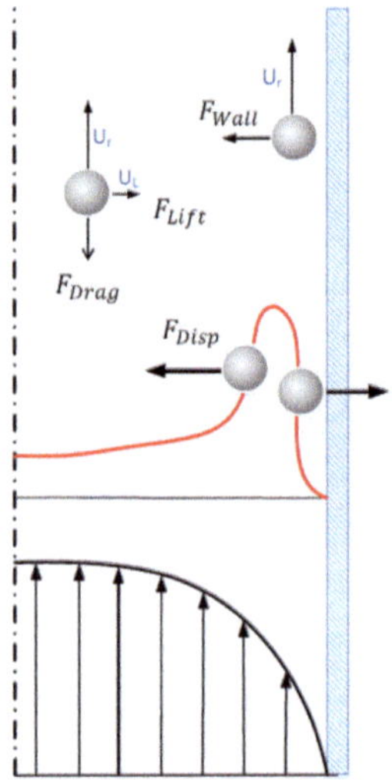

Figure 2.1: Illustration of the interfical forces $\boldsymbol{M}$ in bubbly flows.

Drag Force

The drag force results from the viscous force acting on the bubble surface and the pressure differences caused by the bubble shape (Wang & Yao, 2016). It acts as the resistance between the gas phase and the liquid phase and generates a momentum exchange due to the slip velocity between phases. The corresponding gas phase momentum is defined as

$$\boldsymbol{F}_D = -\frac{3}{4d_p} C_D \rho^L \alpha^G |\boldsymbol{u}^G - \boldsymbol{u}^L| (\boldsymbol{u}^G - \boldsymbol{u}^L)\,, \tag{2.4}$$

with d_p, the bubble diameter. In the following, the terms bubble and particle will be used practically as synonyms, with the index p throughout. Indeed, the term 'particle' in the literature usually designates any element of a disperse phase, such as solid, droplet or gas (Clift, et al., 1978).

The validity of most of the correlations for the drag coefficient C_D is limited to restricted bubble shape regimes (e.g. Schiller & Naumann (1933) for bubble Reynolds number, $Re_p < 800$). A correlation distinguishing different bubble shapes was proposed by Ishii & Zuber (1979), which combines correlations for the spherical, elliptical and cup bubbles regimes. The expression of C_D depends on the bubble Reynolds number, Re_p and the Eötvös number, Eo

$$C_D = max\left(C_{D,Sphere}, \min(C_{D,ellipse}, C_{D,cap})\right), \tag{2.5}$$

where

$$C_{D,Sphere} = \frac{24}{Re_p}\left(1 + 0.1Re_p^{0.75}\right), \tag{2.6}$$

$$C_{D,ellipse} = \frac{2}{3}Eo^{0.5}, \tag{2.7}$$

$$C_{D,cap} = \frac{8}{3}. \tag{2.8}$$

This correlation is based on measurements in contaminated air-water systems and covers a wide range of bubble shape regimes. It seems to fulfil the requirement to model poly-disperse bubbly flows. Tomiyama et al. (1998) validated this correlation and found good agreement except for high values of the Eötvös number.

Lift Force

In shear flows, bubbles experience a force perpendicular to the direction of the relative velocity. This effect generally is referred to as lift force and described by the expression of Auton (1987):

$$\boldsymbol{F}_L = -C_L\rho^L\alpha^G(\boldsymbol{u}^G - \boldsymbol{u}^L) \times rot(\boldsymbol{u}^L). \tag{2.9}$$

For a spherical bubble in simple linear shear flow, the shear-induced lift coefficient C_L is positive. Therefore, the lift force acts in the direction of decreasing liquid velocity, i.e. in case of upwards channel flow in the direction towards the fixed walls.
The situation is complicated for large and non-spherical bubbles. Experimental investigation by Tomiyama et al. (2002) showed that the direction of the lift force changes sign for large bubbles with an ellipsoidal-like shape. This is due to the deformation of bubbles, which was later investigated by Adoua et al. (2009), using DNS for an ellipsoid bubble fixed in a viscous linear shear flow. For prediction of the lift coefficient, Tomiyama et al. (2002) proposed an empirical correlation for C_L based on the observation of bubble trajectories in a Couette flow generated by a moving belt in a tank filled with glycerol and water solution:

$$C_L = \begin{cases} \min\left(0.288\,tanh(0.121Re), f(Eo_\perp)\right), & Eo_\perp < 4 \\ f(Eo_\perp), & 4 < Eo_\perp < 10 \\ -0.27, & Eo_\perp > 10, \end{cases} \tag{2.10}$$

with

$$f(Eo_\perp) = 0.00105Eo_\perp^3 - 0.0159Eo_\perp^2 - 0.0204Eo_\perp + 0.474\,. \tag{2.11}$$

The modified Eötvös number $Eo_\perp$ is given by

$$Eo_\perp = \frac{g(\rho_L - \rho_G)d_\perp^2}{\sigma}, \tag{2.12}$$

where $d_\perp$ is the maximum horizontal dimension of a bubble and can be calculated using the empirical expression suggested by Wellek et al. (1966)

$$d_\perp = d_p\sqrt[3]{1 + 0.163\,Eo^{0.757}}\,. \tag{2.13}$$

Here, $d_\perp$ is related to the normal Eötvös number based on the spherical equivalent bubble diameter. The correlation (2.10) for C_L is illustrated in Figure 2.2, where the lift coefficient changes sign at around $d_p \approx 5.8$ mm.

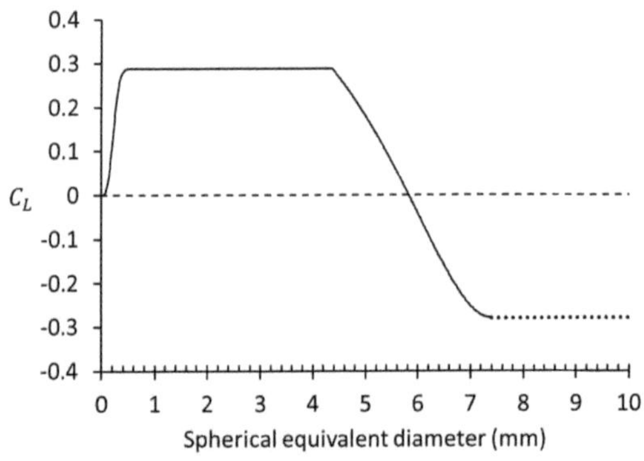

Figure 2.2: Lift coefficient, C_L by Tomiyama et al. (2002) for an air-water system according to Eqs. (2.10) ~ (2.13). The dotted line indicates the region not covered by Tomiyama et al. (2002).

The experiment by Tomiyama et al. (2002) was limited to a modified Eötvös number of around $Eo_\perp \leq 10.7$ for large bubbles, since larger bubbles were not stable. Therefore, the constant value, $C_L = -0.27$ in Figure 2.2 for large bubbles was assumed (region marked with dotted line). The Morten number, *Mo* by their experiment was in the range $-5.5 \leq \log_{10}(Mo) \leq -2.8$. The value is higher than that in the water-air system at normal conditions, with $Mo = 2.63 \cdot 10^{-11}$. Lucas & Tomiyama (2011) investigated an extensive experimental database from bubbly pipe flows and could confirm the applicability of (2.10).

Wall Force

The wall force has been first introduced by Antal et al. (1991), which can be explained as that a bubble ascending close to a wall also experiences a force due to asymmetric drainage of the liquid around it. This force behaves analogous to a lubrication force and acts on a bubble near walls to prevent bubbles from touching the wall (Antal, et al., 1991). The wall force has the general form

$$\boldsymbol{F}_W = \frac{2}{d_p} C_W \rho^L \alpha^G |\boldsymbol{u}^G - \boldsymbol{u}^L|^2 \hat{\mathbf{y}}, \qquad (2.14)$$

with $\hat{\mathbf{y}}$ is the unit normal vector perpendicular to the wall directed towards the fluid. The wall force coefficient C_W depends on the distance to the wall y and is expected to be positive, hence, the bubble is driven away from the wall.
Tomiyama et al. (1995) concluded a correlation for the wall force coefficient based on the experiment of single bubbles rising in the vicinity of a vertical wall. In their experiment, the wall force was evaluated from the bubble trajectories in the mixture of glycerol and water at a constant Morton number ($\log_{10}(Mo) = -2.8$). The expression for C_W is

$$C_W = f(Eo)\left(\frac{d_p}{2y}\right)^2. \qquad (2.15)$$

Hosokawa et al. (2002) experimental confirmed the applicability of (2.15) for a wide range of bubble regime, where $f(Eo)$ is

$$f(Eo) = max\left\{\frac{7}{Re_p^{1.9}}, 0.0217Eo\right\}. \qquad (2.16)$$

The experiment by Hosokawa et al. (2002) was performed under the condition of $2.2 \leq Eo \leq 10$ and $-2.5 \leq \log_{10}(Mo) \leq -6.0$ in the different mixture of glycerol and water, which is still different from the normal water-air system ($Mo = 2.63 \cdot 10^{-11}$). Recent investigation of different wall force correlations has shown that the model of Hosokawa et al. (2002) gives superior results for upward bubbly pipe flow also in air-water system compared to the other models (Wang & Yao, 2016).

Turbulent Dispersion Force

The turbulent dispersion force is the result of the turbulent fluctuations of liquid phase. In dispersed bubbly flows, turbulence in the continuous liquid phase causes bubbles to be transferred from regions of the high concentration to the regions of low concentration (Burns, et al., 2004). In EE-RANS simulations, this contribution has to be modelled, because no turbulence is resolved. Frequently, the model of Burns et al. (2004) derived by Favre averaging of the drag force is used:

$$\boldsymbol{F}_{TD} = -\frac{3}{4} C_D \frac{\alpha^G}{d_p} |\boldsymbol{u}^G - \boldsymbol{u}^L| \frac{\mu_L^t}{\sigma_{TD}} \left(\frac{1}{\alpha^L} + \frac{1}{\alpha^G}\right) \nabla(\alpha^G), \tag{2.17}$$

with C_D and σ_{TD} as the drag coefficient and the turbulent Schmidt number taken to be 0.9, respectively.

2.3 Turbulence modelling

In buoyancy driven bubbly flows, a wide range of length and time scales exists on which turbulent mixing takes place. The largest turbulence scales are comparable in size to the characteristic length of the mean flow and depend on reactor geometry and boundary conditions. The smaller scales depend on the bubble dynamics and hence are proportional to the bubble diameter. In bubbly flows, the small scales are responsible for the dissipation of the turbulent kinetic energy as in single-phase flow, but the bubbles can also generate back-scatter, i.e. energy transfer from smaller to larger scales (Dhotre, et al., 2013). Such a broad spectrum of turbulent scales complicates the treatment of turbulence in the numerical simulation of bubbly flows.
If all turbulence scales, both in space and time, are resolved, the numerical approach is referred to as DNS, without using any turbulence model. Regarding bubbly flows, DNS requires employing interface resolving methods (Tryggvason, et al., 2011).
This is in contrast to the Euler-Euler approach considered in the present study. In principle, it can also be run with a "DNS" mesh size and time step, but the fact that grid points are very close in the Euler-Euler approach does not necessarily mean that the fluid motion is resolved with the same detail. The details have just been removed from the equations and the quantity determined is an averaged quantity. This is the same as solving the equations of the low-Reynolds k-ε model on a grid with step size finer than the Kolmogorov scale which provides the almost exact solution to the RANS problem, but not a DNS. See also the discussion in Geurts & Fröhlich (2002), where the issue is developed in detail for LES. The same applies as well to the point-particle approach, occasionally also addressed as "DNS" of a multiphase flow.

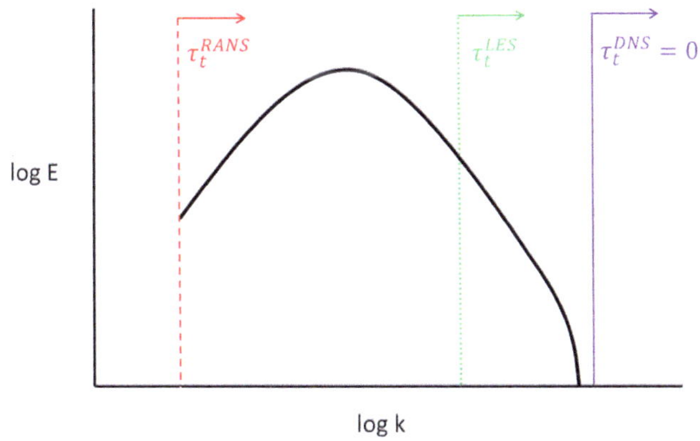

Figure 2.3: A schemmatic diagramm of a prossible energy spectrum with vertical lines illustrating the extent of modelling employed by the numerical simulation strategies of DNS, LES, and RANS.

For the reason mentioned above, in the Euler-Euler approach, SRS and RANS are typically used for the liquid phase, while turbulence inside the bubbles can be safely neglected (Ishii & Hibiki, 2011).

Reynolds-Averaged Navier-Stokes (RANS) Modelling

Two-phase RANS models are formulated on the basis of single-phase RANS modelling using the Reynolds averaging concept. For more basic coverage of single-phase turbulence modelling, refer to books such as Wilcox (1998) and Pope (2000).
The corresponding two-phase RANS equations in the framework of Euler-Euler approach are formulated in terms of the mean flow field and all turbulence is modelled (see Figure 2.3). Similar to single-phase RANS modelling, the focus of two-phase RANS modelling is set on providing a link between the mean flow and the Reynolds stress tensor $\boldsymbol{\tau}_t$ to close the equations. Two-phase RANS modelling is more complicated than single-phase RANS modelling, since the influence of the second phase (e.g. bubbles) has to be taken into account with different approaches introduced in Section 2.4. Details concerning RANS modelling of turbulence modulation in bubbly flows can be found in Section 5.

Scale resolving simulation (SRS)

The type of RANS model typically employed is isotropic and based on an eddy-viscosity, while the turbulence created by some rising bubble swarm (e.g. bubble plume) exhibits large-scale unsteadiness and anisotropy. In principle, all RANS models can be solved in unsteady mode, which then is termed URANS. The basic RANS model, however, is not conceived for partial resolution of the turbulent motion, even if the grid and time step would be small enough, which makes the URANS approach delicate in cases where the unsteadiness is due to an internal instability and not to unsteady boundary conditions, as discussed in Fröhlich & von Terzi (2008).

It is becoming clear that certain classes of bubbly flows are better covered by turbulence models in which all or a part of the turbulence spectrum is resolved (Menter, 2012). In that reference, such methods are termed scale resolving simulations (SRS). These methods are particularly designed to resolve large portions of the turbulent motion and hence seem to be more appropriate for bubble columns than traditional URANS, if large-scale unsteadiness is present. An overview of various approaches to SRS is provided by Fröhlich & von Terzi (2008), together with a classification scheme. LES (see Figure 2.3) falls into the category and offers the opportunity to resolve the large-scale anisotropic turbulence directly and to model the small scales with a subgrid-scale (SGS) model. The scale adaptive simulation (SAS) approach (Menter & Egorov, 2010) is another type of SRS models, which is based on a transport equation for the turbulent length scale using the local von Karman length-scale as a reference. The information provided by the latter allows the SAS model to adjust to resolved structures in an unsteady simulation, so that generally less computational effort is required compared to LES.

In the present study, LES and SAS in unsteady mode are performed for different types of bubbly flows in Sections 3 and 4, respectively.

2.4 Bubble-induced turbulence

In the Euler-Euler approach bubbles are not resolved. Even if in EE-SRS, the resolved part of the velocity field represents only the shear-induced turbulence, which is assumed to be independent of the relative motion of bubbles and liquid. The impact of the bubbles traveling through the liquid (possibly exhibiting bubble wake instability, pseudo-turbulence due to liquid displacement, and bubble oscillations etc.) onto the fluid turbulence has to be modelled.

Algebraic viscosity

Sato & Sekoguchi (1975) and later Sato et al. (1981) suggested to determine the effective viscosity in bubbly flows using a linear superposition of different contributions:

$$\nu_{eff}^{L} = \nu_{mol}^{L} + \nu_{t}^{L} + \nu_{bub}^{L} \,. \tag{2.18}$$

Here, the second component ν_t^L in (2.19) concerns eddy viscosity in the usual sense due to the “true” turbulence (Lance & Bataille, 1991) and the additional eddy viscosity due to bubbles ν_{bub}^L is expressed as:

$$\nu_{bub}^{L} = C_B \alpha^G d_p |\boldsymbol{u}^G - \boldsymbol{u}^L| \,, \tag{2.19}$$

with constant C_B being a model parameter equal to 0.6 (Sato, et al., 1981). This component (ν_{bub}^L) presents the perturbations induced by bubbles, taking account of "drift" phenomena of liquid due to liquid displacement by the bubbles (Sato, et al., 1981). These perturbations cannot be interpreted as turbulence in the conventional sense and are commonly called "pseudo-turbulence" (Lance & Bataille, 1991). However, this notion is sometimes also used for general velocity fluctuations including the wake contributions in the flow, which is solely driven by rising bubbles (e.g. Martinez Mercado et al. (2010); Roghair et al. (2011a)).
Another algebraic modelling of BIT is due to Biesheuvel & van Wijngaarden (1984), which is based on the assumption of a potential flow that the influence of the bubbles on the liquid velocity fluctuations results primarily from the displacement of liquid by the moving bubbles with high Reynold number. The liquid Reynolds stress in their model is written as,

$$\langle \overline{u_i' u_j'} \rangle = \begin{pmatrix} 1/5 & 0 & 0 \\ 0 & 3/20 & 0 \\ 0 & 0 & 3/20 \end{pmatrix} \alpha^G |\boldsymbol{u}^G - \boldsymbol{u}^L|^2 , \tag{2.20}$$

However, such an assumption is not valid for real situations at finite bubble Reynolds number, where the wake contribution is dominant (Risso & Ellingsen, 2002). A simple example is given by Akbar et al. (2012), who used equation (2.20) for pseudo-turbulence to estimate the liquid velocity fluctuations for a homogeneous dense bubble swarms with bubble Reynolds number, $Re_p \sim 1000$. It was found that this part of the contribution was negligible compared to the measured velocity fluctuations, where in such a case the wake contribution was much larger.

BIT Source terms in transport equations

This is the most frequently used approach for considering BIT in the framework of EE-RANS. Additional source terms have been developed for two-equation RANS models (e.g. *k*-*ε*; *k*-*ω*) to describe bubble-induced turbulence. The approximation is provided by the assumption that energy lost by the bubbles due to drag is converted to turbulent kinetic energy in the wake of the bubble. Detailed information about the BIT models in RANS can be found in Section 5. Such an approach is suitable for a turbulence model with transport equation for the turbulent kinetic energy or the Reynolds stress.

Riboux's model

Riboux et al. (2013) reported that the BIT resulting from the wake is controlled by Re_p, $\boldsymbol{F}_D$, and the oscillation frequency of the wake. They reproduced the individual bubble wake by imposing a localized momentum forcing depending on these three main parameters in a single-phase simulation without resolving bubbles (Figure 2.4). The bubbles in this approach can be understood as fixed momentum sources of finite size randomly distributed in the considered flow (Riboux, et al., 2013).

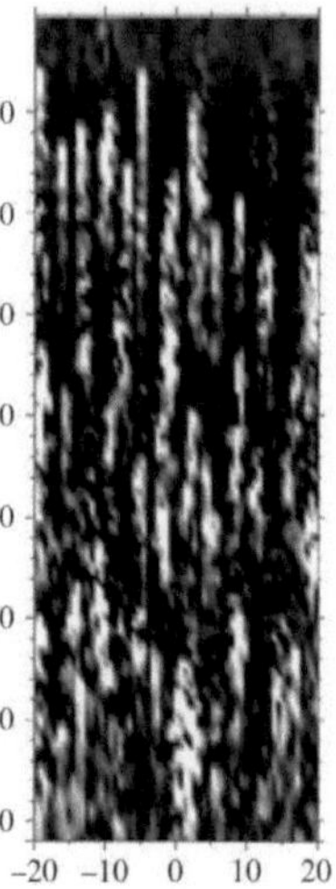

Figure 2.4: Vertical velocity fluctuations by randomly distributed momentum scource, representing many bubbles (Riboux, et al., 2013). [Figure reproduced with permission from Cambridge University Press]

3 Large eddy simulation for dispersed bubbly flows

Bubble columns provide a good experimental system for the study of turbulent phenomena in bubbly flows and a suitable setup for the development of computational models. In bubble columns a wide range of length and time scales exists on which turbulent mixing takes place. In the CFD simulations of bubble columns, RANS models are used for modelling turbulence in the traditional way, using isotropic closures without resolution of turbulent scales. However, in some gravity-driven bubbly flows, a distinct transient behaviour can be identified through large-scale circulation as reviewed by Mudde (2005). Also, through the uneven aeration, naturally caused by the sparger in larger bubble columns, a distinct periodicity of the flow occurs, which is studied for example by Pfleger et al. (1999) and Julia et al. (2007). Therefore, an influence of the transient processes can be assumed which a usual steady-state RANS model is not able to capture.

The computation of such flows in a bubble column poses severe challenges to turbulence modelling due to the presence of dominating anisotropic large-scale vortex systems. Under these circumstances, LES is an approach potentially capable of yielding improved results compared to RANS methods due to the reduced amount of modelling which is traded against computational resolution and hence CPU time (Fröhlich, et al., 2002). On the other hand, with the Euler-Euler approach the mesh requirement in such an LES is comparable to unsteady RANS simulations as demonstrated e.g. by Deen et al. (2001) and Dhotre et al. (2008).

In the present section the performance of LES employing the Euler-Euler approach for such kind of flows is investigated. One of the challenges in such an approach is to deal with a wide range of length and time scales involved in bubbly flows, which even a challenge still exists in LES for single-phase turbulent flows (Fröhlich, 2006). Regarding to bubbly flows, additional difficulties arise due to the small-scale turbulence induced by the dispersed gas phase, which are further magnified the challenge mentioned above (Fox, 2012).

Simulations related to EE-LES have been performed by different authors. Zhang et al. (2006) used LES with the Smagorinsky model to simulate a square cross-sectional bubble column, with the gas inlet placed in the centre of the bottom. They compared the results obtained with different values of the Smagorinsky constant C_S and found that too high values lead to an unphysically high effective viscosity which in turn damps the bubble plume dynamics. Such an effect should not be found for bubble columns with a gas inlet at the bottom that homogeneously distributes the bubbles over the entire cross section, because the gas volume fraction profile in such bubble columns is flat and the impact of the increased effective viscosity is lower (Ma, et al., 2015a). Dhotre et al. (2008) reported LES with a two-fluid model for the same experiment as by Zhang et al. (2006). They investigated the influence of SGS models in LES using the Smagorinsky model and the dynamic Smagorinsky model of Germano et al. (1991). It was found that the performance of both models was similar. In fact, the averaged value of C_S in the dynamic Smagorinsky model was close to the value of C_S in the Smagorinsky model. The results obtained compare well with the experiments. Furthermore, an extensive discussion of the merits of LES can be found in the work of Dhotre et al. (2013), providing a systematic evaluation of prior work on the modelling of turbulent bubbly flows.

This chapter consists of three parts. After a brief introduction in Section 3.1 for the method of EE-LES employed, the second part regarding Section 3.2, deals with a bubble column with homogeneously distributed gas inlet at the bottom for two gas inlet velocities. EE-LES is employed to investigate its performance in different flow region, i.e. with or without large-scale turbulent structure. Conclusions are finally presented at the end of the section. The third part, Section 3.3, focuses on a bubble plume and follows the work of Pfleger et al. (1999) using a 3D URANS simulation, which successfully predicted the time-averaged liquid vertical velocity of different heights of the column. However, difficulties arise with their approach to reproduce the amplitude and the large-scale period of the horizontal fluid velocity in the experiment. In contrast to their work, here, EE-LES is performed for the same bubble column to improve the results. The results of this chapter have been published in Ma et al. (2015a) and Ma et al. (2016).

3.1 Description of method employed

Equations (2.1) and (2.2) are usually derived by ensemble averaging. However, the same form of the equations is obtained if one performs filtering (volume averaging) of the governing equations (Niceno, et al., 2008). This is of practical importance for LES, because it means that the same numerical tools developed for ensemble averaged Euler-Euler equations, can be used for LES. The difference then only resides in the model term $\boldsymbol{\tau}_t$.
In this study, turbulence is treated differently for the two phases. The turbulence in the dispersed gas phase is of little relevance and is modelled with a simple zero equation model $\mu_t^G = (\rho^G/\rho^L)\mu_t^L$ (Ansys, 2013). It was found that this model has nearly no influence on the result (Ma, et al., 2015b).

Subgrid-scale modelling

The liquid velocity $\overline{\boldsymbol{u}}^L$ in (2.1) and (2.2) represents the resolved velocity contribution. The index L is dropped in the remainder of this thesis for better readability, unless stated otherwise. The corresponding unresolved contributions are:

$$\boldsymbol{u}' = \boldsymbol{u} - \overline{\boldsymbol{u}}\,, \tag{3.1}$$

with $\boldsymbol{u}$ the true velocity of liquid. Two SGS models of the eddy viscosity type are used in the EE-LES below. In both models, the anisotropic part of the SGS stress tensor $\tau_{ij} = \overline{u_i u_j} - \bar{u}_i\bar{u}_j$, is modelled as

$$\tau_{ij}^a = \tau_{ij} - \frac{1}{3}\tau_{kk}\delta_{ij} = -2\nu_{sgs}\bar{S}_{ij} \tag{3.2}$$

with τ_{ij}^{a} being the anisotropic (traceless) part of the SGS stress tensor τ_{ij}, and δ_{ij} the Kronecker delta. The two SGS models differ in determining the SGS viscosity (or eddy viscosity) ν_{sgs}.

For the simulation of bubble column in Section 3.2, the Smagorinsky model (Smagorinsky, 1963):

$$\nu_{sgs} = (C_s\Delta)^2|\bar{S}|\,, \tag{3.3}$$

is applied, with the SGS viscosity, ν_{sgs}, being a function of the magnitude of the strain rate tensor, $|\bar{S}| = \sqrt{2\bar{S}_{ij}\bar{S}_{ij}}$, and the subgrid length scale is $l = C_s\Delta$. Here, the model constant was chosen to be $C_s = 0.15$ based on previous experience of Akbar et al. (2012), while the filter width Δ was determined by the grid size according to $\Delta = \sqrt[3]{\Delta x \Delta y \Delta z}$, which is convenient with the cartesian grid employed for the simulation as described below. Close to the walls the turbulent viscosity is damped using the formulation of Shur et al. (2008). The trace of the SGS stress tensor τ_{kk} in (3.2) is added to the filtered pressure $\bar{p}$, resulting in a modified pressure $P = \bar{p} + \frac{\tau_{kk}}{3}$. For wall modelling, a blending is performed between the viscous sublayer and the logarithmic law of the wall depending on the mesh resolution (Ansys, 2013).

The second model used in Section 3.3 is the dynamic Smagorinsky model of Germano et al. (1991), with the modification of Lilly (1992). The model coefficient C in the Smagorinsky expression

$$\nu_{sgs} = C\Delta^2|\bar{S}|\,, \tag{3.4}$$

is obtained in the dynamic procedure using an explicit box filter of width twice the grid size. To avoid numerical instability, a relaxation of C in time is applied and an upper and lower limit of the coefficient is imposed with $C^{max} = 0.04$ and $C^{min} = 0$.

Bubble-induced turbulence in EE-LES

With the Euler-Euler approach bubbles are not resolved. The resolved part of the velocity field in LES represents only the shear-induced turbulence, which is assumed to be independent of the relative motion of bubbles and liquid. The impact of the bubbles traveling through the liquid onto the fluid turbulence has to be modelled. In two-equation RANS models, additional source terms have been developed to describe bubble-induced turbulence. For EE-LES, Niceno et al. (2008) employed the one-equation SGS model of Davidson (1997) with an extra source term developed by Pfleger & Becker (2001) in the transport equation to represent the effect of BIT and computed a square cross-sectional bubble column. It was found that this approach gives good predictions (Niceno, et al., 2008). However, the unresolved part of the turbulent kinetic energy reaches about 20% with this approach, which might be too high for a traditional LES (Fröhlich & von Terzi, 2008).

However, such an approach is not suitable for LES with a zero equation SGS model, because no transport equation for the turbulent kinetic energy k is available. Here, the common BIT model of Sato et al. (1981) is used. In this model the bubble influence on liquid turbulence is included by an additional extra term contributing to the SGS turbulent viscosity (see equations (2.18) and (2.19)). In these EE-LES, ν_{bub} is added directly in the SGS model, without direct contribution to the total turbulent kinetic energy, i.e.

$$\tau_{ij}^{a} = \left(\tau_{ij} - \frac{1}{3}\tau_{kk}\,\delta_{ij}\right) = -2(\nu_{sgs} + \nu_{bub})S_{ij}\,, \tag{3.5}$$

since τ_{kk} is not computed anyway.

Interfacial forces

The corresponding interfacial transfer models employed here are listed in Table 3.1. A description of the employed interfacial transfer models can be found in Section 2.2. Here, the turbulent dispersion model is not employed with EE-LES. Since the resolved part of the turbulent dispersion in EE-LES can be calculated explicitly and the unresolved part has little influence on bubble dispersion if the bubble size is in the scale of the filter size (Niceno, et al., 2008).

Drag force	Ishii & Zuber (1979)
Lift force	Tomiyama et al. (2002)
Wall force	Hosokawa et al. (2002)

Table 3.1: Models for interfacial forces employed for all cases in the present chapter.

3.2 Bubble column with homogeneous distributed gas inlet

3.2.1 Experimental data and simulation setup

The simulations were carried out for a rectangular water/air bubble column at ambient pressure (Akbar, et al., 2012) and are compared here with the experimental data for two different gas superficial velocities at the inlet. A schematic sketch of the experimental setup is shown in Figure 3.1. Its width, depth and height are 240, 72 and 800 mm, respectively, and the water level is 700 mm. A distributor plate containing 35 evenly spaced needles with an inner diameter of 0.51 mm was positioned at the bottom of the experimental column. Measurements using a laser Doppler velocimetry (LDV) system, an electrical conductivity probe and a high speed camera were performed for two superficial velocities 3 mm/s (Case 1) and 13 mm/s (Case 2). Liquid velocities, void fraction and bubble velocities were measured along the line 500 mm above distributor plate in the centre plane (y = 36 mm). More details are provided in the cited reference (Akbar, et al., 2012).

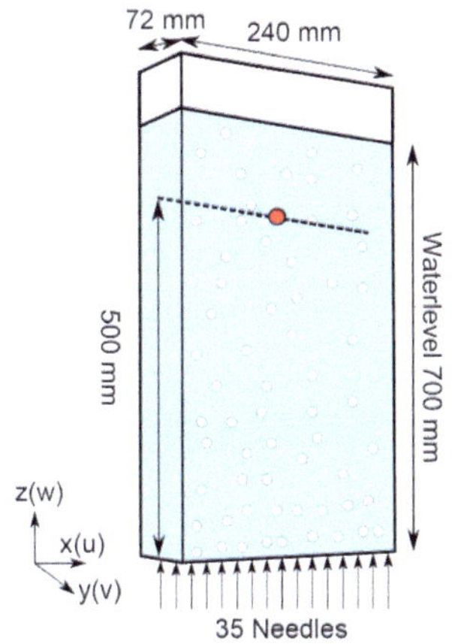

Figure 3.1: Schematic representation of the experiment of Akbar et al. (2012). The broken line in the figure shows the measurement position. The marker on that line is the measurement point for the results presented in Figure 3.9 and Figure 3.10.

The measured bubble size distributions at the inlet and measurement plane are given in Figure. 3.2. Case 1 exhibits a relatively narrow distribution, while the distribution of Case 2 is much wider. Obviously, the bubble size distribution near the sparger and 500 mm above the sparger is almost the same for both cases. Therefore, coalescence and break-up will be neglected. For modelling the polydispersity, the inhomogeneous Multiple Size Group (iMUSIG) model as introduced by Krepper et al. (2008) assigns different velocity groups to the bubble classes used in the MUSIG model. It is based on bubble size classes for the mass balance as well as for the momentum balance. Each velocity group has therefore its own velocity field. This is important, to describe effects like the bubble size-dependent movement of the gas phase caused by the lift force. In the present case, the bubble classes are chosen in such a way that the bubble size distributions are split into two contributions at the diameter where the lift force changes its sign, which is $d_p = 5.8$ mm. The resulting bubble classes for Case 2 can be found in Table 3.2. In Case 1 bubbles are treated as monodisperse with $d_p = 4.37$ mm, because almost all bubbles have a positive lift coefficient, so that there is no need for considering different velocity groups.

	d_p	$\propto$	$Eo_\perp$	C_L
Bubble Class 1	5.3 mm	0.63	3	0.288
Bubble Class 2	6.3 mm	0.37	7.3	-0.116

Table 3.2: Bubble classes employed in Case 2. $\propto$ is the corresponding ratio in total gas.

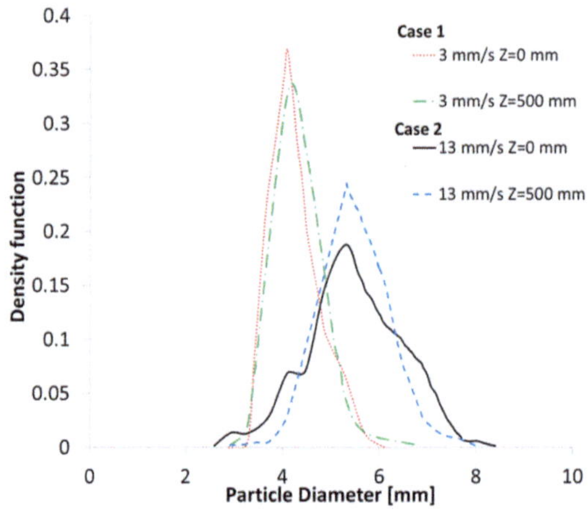

Figure. 3.2: Measured bubble size distribution at z = 0 and 500 mm (Akbar, et al., 2012).

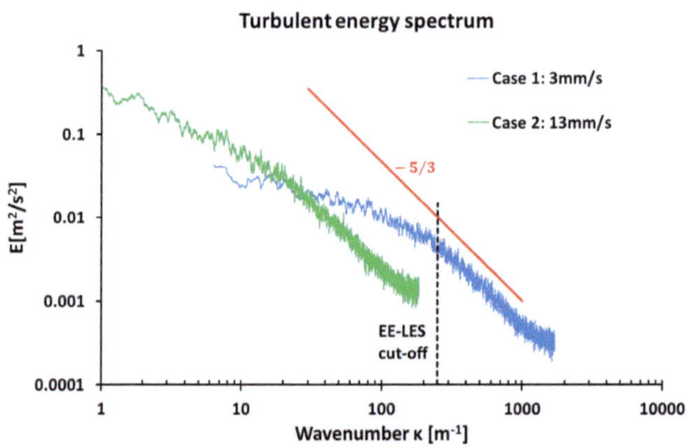

Figure 3.3: Turbulent energy spectrum of fluctuating vertical velocity.

Grid Requirements

The rectangular bubble column was discretized with uniform cubic cells of $\Delta x = \Delta y = \Delta z =$ 4 mm, resulting in about 200000 cells, overall. The spatial resolution is in the order of the bubble size ($d^* = \Delta x/d_p \sim 1$). Resolution requirements for Euler-Euler methods are still subject of debate in the literature. The approach was conceived to represent bubble swarms on very coarse grids, far coarser than individual bubbles. The equations are derived by means of volume averaging, and the averaging volume then has to be substantial larger than individual bubbles. Once the equations derived, the problem to be solved is composed of a partial differential equation (PDE) and appropriate boundary conditions. This continuous problem can be solved numerically on an arbitrarily fine grid, then yielding a very smooth solution. Indeed, the numerical solution converges to the exact solution of the Euler-Euler PDE, so that the discretization error is negligible. The modelling error persists, so that caution is needed only when interpreting the result. In the present work, the grid size was chosen according to the experimental energy spectrum. Figure 3.3 shows the energy spectrum of the vertical velocity calculated at the measurement point in Figure 3.1 using a Fast Fourier Transformation (FFT). The energy spectrum from Akbar et al. (2012a) was transformed from frequency in time to wavenumber in space using Taylor's hypothesis of frozen turbulence. This allows to insert the LES cut-off (filter size = 4 mm) into the energy spectrum, as done in Figure 3 with the dotted line. For Case 2 the grid is fine enough and for Case 1 the grid scale is located roughly at the beginning of the inertial subrange, which is coarser than what usually would be chosen. However, as will be reported in Section 3.2.3 below, the vertical velocity fluctuations are in the order of the vertical mean flow, so the condition $w'/\overline{w} \ll 1$ for using Taylor's hypothesis is actually not fulfilled. Figure 3.3, nevertheless, provides some hint concerning the resolution.

Simulation Details

The needles introducing the bubbles in the bottom plate of the container are modelled by small $4\ mm \times 4\ mm$ surfaces at the respective locations, each representing one needle. Here, the inlet velocity is imposed to equal the gas velocity, the volume fraction set to 1 for the gas and 0 for the liquid. The gas velocity is obtained by dividing the total gas flux measured in the experiment by the number of needles, so that it is assumed to be equal at all needles. The liquid velocity at the inlet is set to 0. For the initial conditions, the velocities are set to 0 for both phases and the volume fraction defined as: $\alpha^L = 1$ and $\alpha^G = 0$, since the liquid is stagnant and all cells of the grid are completely filled with liquid at the beginning. During the entire calculation, a minimal volume fraction 10^{-10} for each phase is set for numerical robustness, so that all volume fractions are larger than 0 all over the computational domain. At the walls, a no slip condition is applied for the continuous phase and a free slip condition for the dispersed phase. At the top of the column a degassing boundary is imposed, which means an outlet for the dispersed gas phase. Here, the outlet pressure is not specified. Instead, a pressure distribution is computed on this fixed-position boundary, and can be interpreted as representing the weight of the surface height variations in the real flow. The continuous liquid phase sees the boundary as a free-slip wall and does not leave the domain. For the spatial discretization, a central difference scheme is employed, and a second order backward Euler scheme is used in time. The simulations were conducted with the code CFX 14.5 and carried out using time steps $\Delta t = 0.01 \cdots 0.015\ s$ to satisfy CFL < 1. The results were averaged over 250 s physical time.

3.2.2 Transient results

To provide a first impression of the EE-LES simulation result, Figure 3.4 presents selected instantaneous data from Case 2 at an arbitrary instant and compares them to the same data from the EE-URANS. The right graphs depict the corresponding void fraction, showing a clear difference between the results for the two turbulence models. The computed streamlines of liquid velocity in Figure 3.4 illustrate that the LES resolves much more of the details of the flow field. With URANS, the transient details are not well resolved. It can be seen that only some large-scale fluctuations are obtained, while smaller ones are damped due to the high turbulent viscosity. Similar differences of the liquid velocity field with LES and URANS are also described by Deen et al. (2001) and Dhotre et al. (2008). This difference in the liquid velocity field yields a more homogenous gas volume fraction distribution in the URANS simulation.

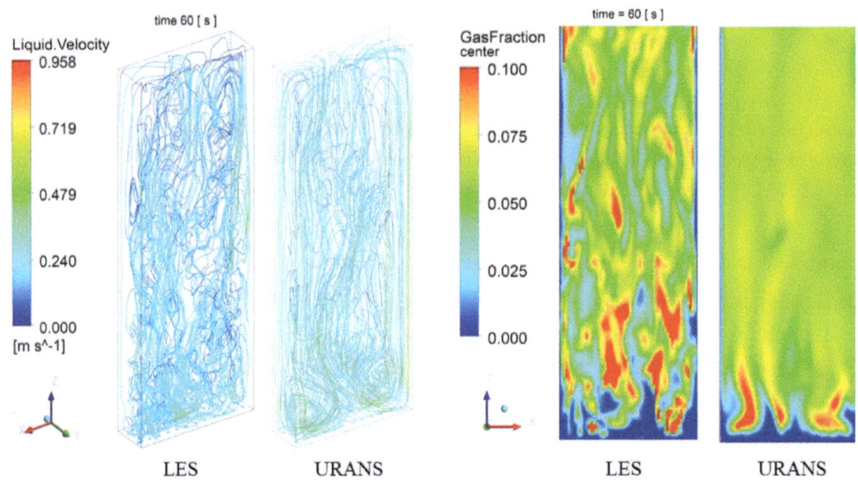

Figure 3.4: Instantaneous data from Case 2 at $t = 60\ s$. The left two graphs show instantaneous streamlines of the liquid colored with the instantaneous absolute value of the liquid velocity. The right two graphs show the instantaneous void fraction in the centre plane.

3.2.3 Time-averaged results

In this section, the results from EE-LES and EE-URANS are compared with previous work of Akbar et al. (2012), who used a Lagrangian modelling for bubbles i.e. EL-LES. The interfacial force models used in his reference are comparable to the settings defined in Section 2.2, and the simulation was performed using two way coupling.

In Figure 3.5 and Figure 3.6, long-time averaged vertical liquid velocity, gas volume fraction and the fluctuations of the vertical liquid velocity are presented for both cases, together with experimental data of Akbar et al. (2012) and EL-LES data from the same reference. All simulations presented were run for 250 s. All profiles were taken along the measurement line from the wall to the centre at a height of 500 mm, as represented by the broken line in Figure 3.1.

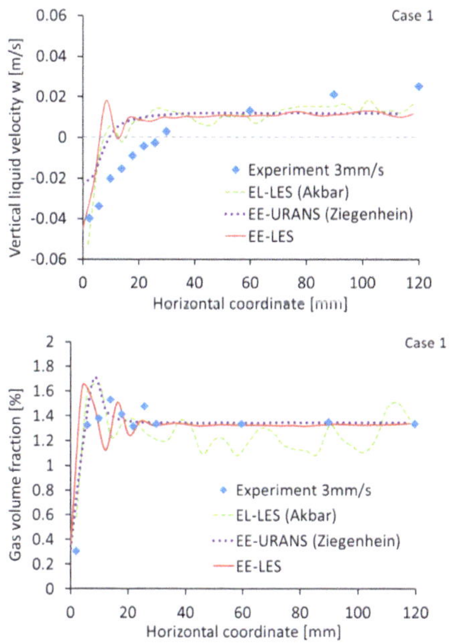

Figure 3.5: Comparison of vertical liquid velocity (top) and gas volume fraction (bottom) for Case 1. Experimental data and EL-LES are from Akbar et al. (2012). EE-URANS are from Ziegenhein et al. (2015).

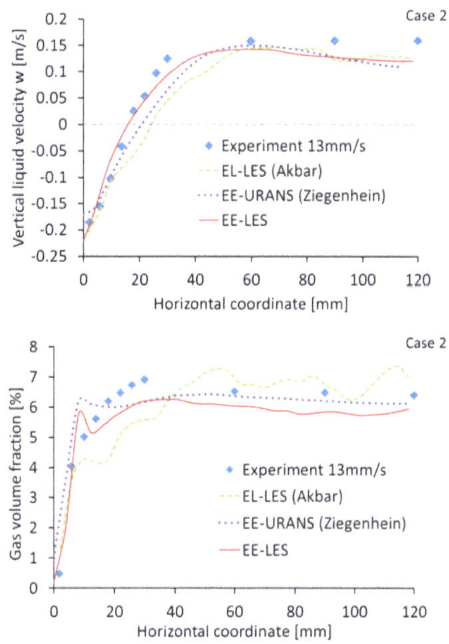

Figure 3.6: Comparison of vertical liquid velocity (top) and gas volume fraction (bottom) for Case 2. Experimental data and EL-LES are from Akbar et al. (2012). EE-URANS are from Ziegenhein et al. (2015).

Figure 3.5 shows the vertical liquid velocity and gas volume fraction for Case 1. The three predicted liquid velocity profiles provide too high velocities near the wall and somewhat too small velocities in the centre. Undesired also is the oscillatory behaviour of the EE-LES near the wall. This can be traced back to an uneven distribution of the void fraction seen in the second picture of that figure. Indeed, also the experimental data exhibit an undulating shape. But the amplitude is smaller and the region of occurrence somewhat further from the wall. So, apparently, this is a feature of this case and while not captured quantitatively by the EE-LES, qualitatively correct. The gas fraction obtained with the other two approaches also exhibits a local maximum, but in the URANS case no and in the EL-LES only small undulations of the velocity data. The wall peak in the void fraction profile is caused by the lift force modelling and is also described by Krepper et al. (2007), who considered a similar experimental setup. The gas volume fraction of about 1.2 % in the centre is obtained in all the simulations and the experimental data. The results from Akbar et al. (2012) using EL-LES exhibits small oscillations of the averaged gas volume fraction and vertical liquid velocity in all results reported. The reason could be a somewhat too small averaging time.

In Figure 3.6, the results for high superficial gas velocity at the inlet (Case 2) are presented. A clear change in the direction of the liquid velocity can be seen at about 20 mm away from wall. This is a phenomenon caused by the liquid mass balance in the bubble column, obtained in all three predictions with a quite good quantitative agreement. In the near wall region the velocity data obtained with EE-LES match the experimental data better than the other two simulations. The gas volume fraction profile in Figure 3.6 is smoother than in Case 1. The experimental data only exhibit a slight overshoot between 30 and 60 mm from the wall. The simulation data show the same tendency of creating a local maximum as observed with Case 1, albeit much less pronounced. This peak caused by the lift force modelling making the small bubbles go towards to the wall, and migrate the big bubbles to the centre, results in a slight second peak in the position about 40 mm away from the wall. The same phenomenon appears also in the URANS simulation using the same lift force modelling. The gas volume fraction results obtained with EL-LES underpredict the void fraction in the wall region and generally exhibit a very jaggy profile, presumably due to lack of averaging.

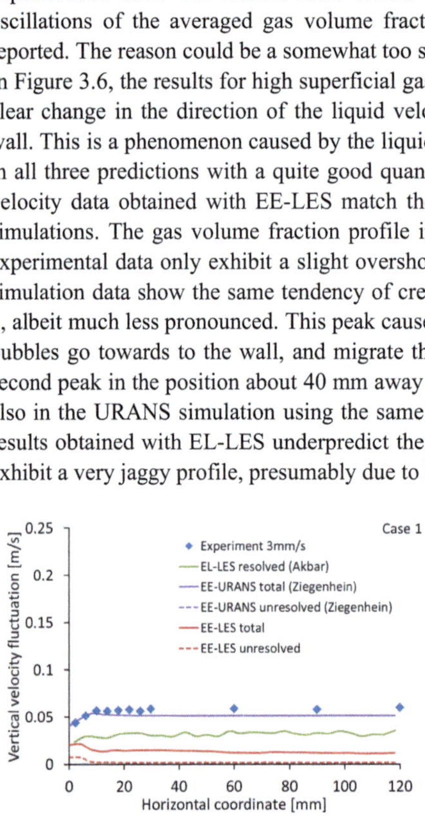

Figure 3.7: Comparison of vertical liquid velocity fluctuation for Case 1. Experimental data and EL-LES are from Akbar et al. (2012). EE-URANS are from Ziegenhein et al. (2015).

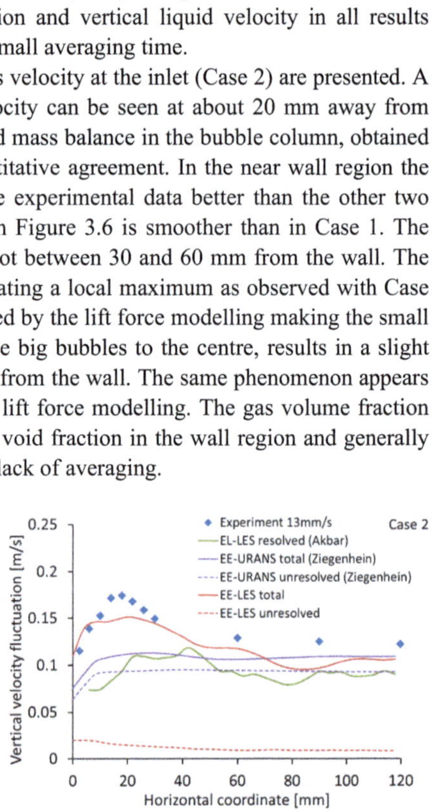

Figure 3.8: Comparison of vertical liquid velocity fluctuation for Case 2. Experimental data and EL-LES are from Akbar et al. (2012). EE-URANS are from Ziegenhein et al. (2015).

Figure 3.7 and Figure 3.8 show comparisons between experimental and numerically predicted vertical liquid velocity fluctuations for Case 1 and Case 2, respectively. With the EL-LES only the resolved velocity fluctuation is provided in Akbar et al. (2012), the unresolved part is neglected. In the EE-LES the total vertical velocity fluctuation is considered, which can be decomposed into the resolved part $\overline{w}''$ and the unresolved part w' as mentioned in Section 3.1. For the resolved part, $\overline{w}''\overline{w}'' = [\overline{w} - \langle\overline{w}\rangle]^2$ is computed using the difference between the resolved vertical velocity and the time average of the resolved vertical velocity. For the unresolved part, as can be seen in (3.2), only the anisotropic part of the SGS stress tensor is considered in the Smagorinsky model, so that the information about the isotropic part $\tau_{kk} = 2k$ is lost. Dhotre et al. (2013) performed a review about application of LES to dispersed bubbly flows, observing that in all papers the collected SGS kinetic energy $\mathrm{k_{sgs}}$ was neglected when zero equation models were used in LES. Only Niceno et al. (2008) demonstrated the applicability of a one-equation model for $\mathrm{k_{sgs}}$, so that $\mathrm{k_{sgs}}$ could be explicitly calculated. Here, a method for estimating $\mathrm{k_{sgs}}$ will be introduced based on the SGS dissipation ε_{sgs} as proposed by Menter (2013) for single-phase flows:

$$k_{sgs} = \sqrt{\frac{\nu_{sgs}\varepsilon_{sgs}}{C_\mu}}, \quad \varepsilon_{sgs} = \nu_{sgs}\,|\bar{S}|^2\,, \tag{3.6}$$

with constant $C_\mu = 0.09$ taken from the standard *k*-ε model. As can be seen in Figure 3.7, for Case 1 both LES with Euler-Euler and Euler-Lagrangian obtain much lower values for the vertical velocity fluctuations than the experimental data. The unresolved part from EE-LES is quite small and has nearly no influence on the evaluation. The underprediction can be related to the limitation of LES for this kind of flow, since large-scale turbulence is not present with a homogeneous distributed low gas inlet at the column bottom. With the EE-URANS the experimental data can be reproduced very well. The total and the unresolved vertical fluctuation profiles are nearly the same, so that the resolved fluctuations using the EE-URANS method are zero for Case 1. Hence, all the velocity fluctuations come from the used two-equation turbulence model with the BIT model. This is in line with the concept of a RANS model and recovered for URANS as well if instabilities in the flow are small. Summarizing, this might be a hint that the large-scale fluctuations are in general very low in Case 1 and the bubble-induced turbulence is dominant. Because the bubbly flow subgrid models used for the EE-LES method include only dissipation terms, as mentioned in Section 3.1, the total velocity fluctuations with this method might be underpredicted.

Figure 3.8 shows the same velocity fluctuations for Case 2. In this case, higher large-scale turbulence is expected, because of the higher superficial velocity. The result of the EE-LES has a better quantitative agreement with the measured data than the other two simulations shown in Figure 3.8. Especially the measured peak is only reproduced by the EE-LES and located at the change of sign of the vertical liquid velocity, as mentioned in Figure 3.6. The unresolved velocity fluctuations obtained by using the EE-LES method in Figure 3.8 amount to about 10% of the total velocity fluctuations. The result with EL-LES underpredicts the fluctuation in this case, which might be caused by neglecting the unresolved part. The trend of a peak close to the wall can also be found in the resolved part of EL-LES.

The results from EE-URANS fit the measured profile well in in Case 1 (Figure 3.7), while the agreement is less satisfactory in Case 2 (Figure 3.8). URANS with a two-equation turbulence model for prediction of velocity fluctuation in one direction might be critical, because the isotropic assumption of turbulence leads to $\overline{w'w'} = \overline{v'v'} = \overline{u'u'} = \frac{2}{3}k$. Similar to Case 1, the unresolved part in Case 2 contributes almost 90% of the total fluctuation (Figure 3.8). So, the lateral velocity fluctuations in the other two directions in both cases would have nearly the same profiles like the vertical direction. Therefore, it might be a problem at this point that only one velocity component is evaluated for the EE-URANS approach. The previous work of Dhotre et al. (2008) also mentions the limit of URANS models to predict the liquid velocity fluctuation in one direction. The comparison between the total turbulent kinetic energy and experimental data could give a better agreement. Unfortunately, the experimental data consists only of one component of the velocity fluctuations.

3.2.4 Analysis of spectra

Figure 3.9 shows a 200 s time history plot of the resolved vertical liquid velocity with 20000 sample points obtained from the EE-LES of Case 2 at the measurement point shown in Figure 3.1.

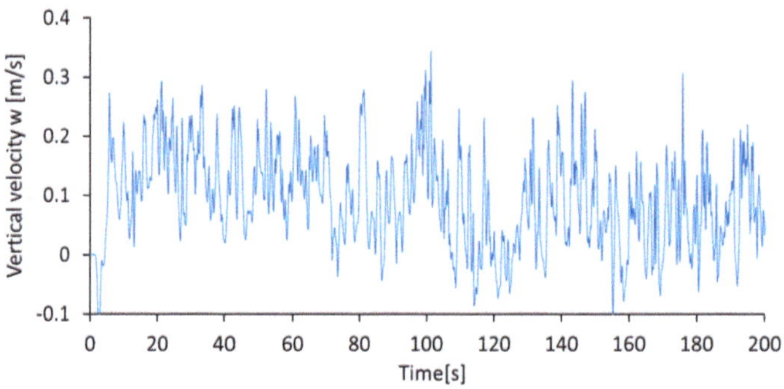

Figure 3.9: Time history of the liquid velocity obtained with EE-LES at the centre of the measurement line for Case 2.

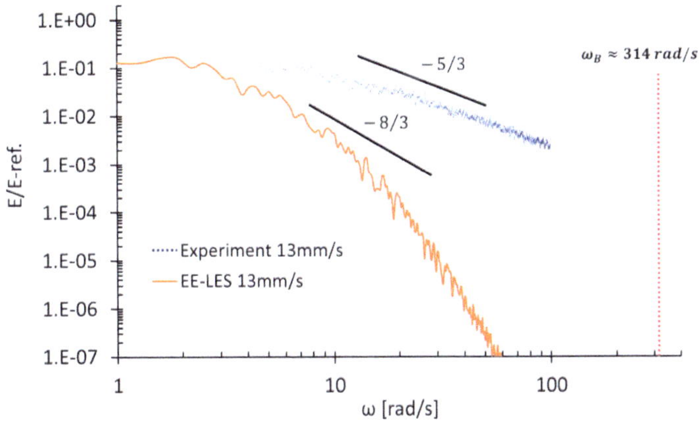

Figure 3.10: Temporal energy spectrum of vertical liquid velocity scaled with an arbitrary reference value. The red dotted line is the estimated frequency of the bubble wake.

The energy spectrum obtained with the data extracted from Figure 3.9 is shown in Figure 3.10. The velocity signal was transformed using the Welch method (Welch, 1967) with 10 non-overlapping windows. Each window contains 2000 sample points, and a Hanning window function was used. As can be seen, the turbulent energy spectrum with the EE-LES approach exhibits a broad range of frequencies, with a slope steeper than the $-8/3$ power law in the inertial subrange, which is still in discussion for this type of flow (Mercado, et al., 2010). Here, the steep slope in the simulation is mainly caused by the BIT model of Sato et al. (1981), which induced an additional enhancement of the eddy viscosity in (3.5). This is a similar effect as increasing the Smagorinsky constant C_s in the LES, which is discussed by Fröhlich (2006). A similar energy spectrum based on EE-LES results using the Sato model was obtained by Dhotre et al. (2008). In his work the slope was partly even over $-10/3$ in the inertial subrange.

Previous experimental studies have actually attributed the more dissipative spectrum to the presence of the bubbles suggesting a mechanism of eddy disintegration (Lance & Bataille, 1991). But in the experimental energy spectrum for Case 2 of Akbar et al. (2012) shown in Figure 3.10, such a dissipative slope does not appear. The difference between the EE-LES energy spectrum and the experimental energy spectrum could be caused by many different reasons. A determinant reason might be the limitations caused by the Euler-Euler approach. It might not be able to reproduce a similar power law as the experimental one. Since in the Euler-Euler approach, bubbles are not resolved, it is impossible to catch the frequency information in the bubble wake like in the experiment. Furthermore, the time scale of the bubble wake is also shown in Figure 3.10. The estimation of the angular frequency ω_p for the bubble wake is introduced as: $\omega_p = 2\pi f_p \approx 2\pi\, u_p/d_p \approx 314\, rad/s$, with f_p being the frequency of the bubble wake, $U_p \approx 0.3\, m/s$ the bubble velocity, and $d_p \approx 6\, mm$ as the averaged bubble diameter for Case 2. The resolved and reliable angular frequencies in the spectrum of the EE-LES are far away from the requirement to know bubble wake information (e.g. BIT). Also in the experiment, this requirement is not reached. Nevertheless, the low frequency contribution can be compared.

3.3 Bubble plumes

3.3.1 Experimental data and simulation setup

The present section follows the work of Pfleger et al. (1999). Here, Euler-Euler LES is performed for the same rectangular air/water bubble column (Figure 3.11) at ambient pressure (Pfleger, et al., 1999). The bubble column has a cross-section of $0.2\ m \times 0.05\ m$ and is filled with distilled water up to a height of 0.45 m. One distributor plate containing 8 needles was placed in the centre of the bottom with a cross-section of $A_{in} \approx 0.02\ m \times 0.0125\ m$. Measurements were performed for a gas superficial velocity of 1.7 mm/s and took place 0.13 m, 0.25 m and 0.37 m above the distributor plate in the centre plane (z = 0.025 m). A LDV system was used to simultaneously measure the liquid velocity.

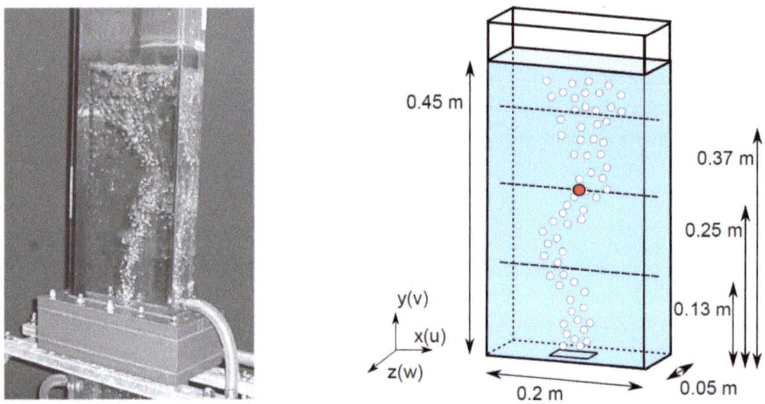

Figure 3.11: The left graph shows the laboratory bubble column of Pfleger et al. (1999). [Picture reproduced with permission]. The right graph shows the schematic representation of this bubble column. The broken lines in the figure show the measurement positions of the three heights of 0.13 m, 0.25 m and 0.37 m from the bottom. The marker on the line at 0.25 m is the measurement point for the 1-D time signal discussed below.

The rectangular bubble column was discretized with uniform cubic cells of $\Delta x = \Delta y = 5mm$ and $\Delta z = 3.125$ mm, resulting in 57600 cells, overall. The mesh was selected here according to a mesh study as discussed in Section 3.3.3 below. The bubbles are treated as mono-disperse with bubble diameter $d_p = 2\ mm$. The gas inlet is defined as a distributed source in the central area $0.02\ m \times 0.0125\ m$ at the bottom of the domain, representing the

experimental needle setup. The boundary and initial conditions are the same as for the case Akbar in Section 3.2.1, except for the localization of the gas inlet.
While no overall mass flux of the fluid is generated due to the boundary conditions, the forcing by the gas phase results in a substantial recirculating flow. An order of magnitude is provided by the Reynolds number built with the averaged liquid velocity at the centre point of the bubble column and the depth of the column, which reaches a value of 5618. The simulations were carried out using a time step $\Delta t = 0.02\ s$ to satisfy $\mathrm{CFL} < 1$. The simulations were conducted with the code CFX 14.5 and were run for a physical time of 500s.

3.3.2 Transient results

In general, the flow structure of a long bubble plume is divided into two distinct regions as shown in Figure 3.12; an oscillating region, where the plume retains its structure and exhibits lateral oscillations, and a non-oscillating region where the plume reaches the walls of the column and the bubbles are distributed, all over the cross-section of the bubble column (Vanga, et al., 2005).

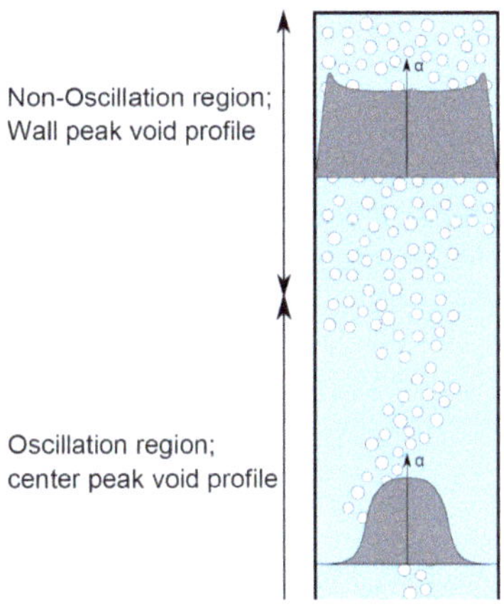

Figure 3.12: two distinct regions of “long” bubble plume and the time averaged void fraction profiles in these regions.

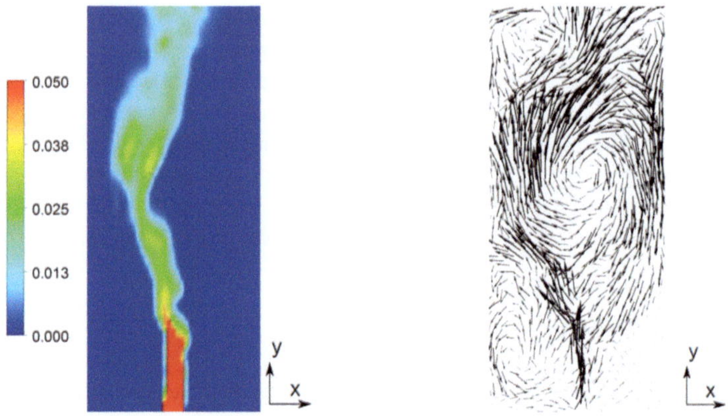

Figure 3.13: The left graph shows shows the instantaneous void fraction in the center plane at t = 50 s. The right graph shows liquid velocity vectors of the corresponding flow field.

Snapshots of the void fraction distribution and the liquid velocity vectors in the centre plane are shown in Figure 3.13 for the simulation using the dynamic Smagorisky without BIT model. These graphs are taken after the statistically stationary state is reached and give good illustrations of the dynamics of the bubble plume. They look very similar to the experimental snapshot in Figure 3.11 and the numerical results presented by Pfleger et al. (1999). Figure 3.11 and Figure 3.13 show that this bubble column exhibits only the oscillating region over the entire height which is characterized by the centre peak void fraction profile and the presence of large-scale vortical structures. The vortex has a length of the order of half the width of the bubble column. As a matter of fact, the bubble column is not long enough to reach the non-oscillating region mentioned above and the flow is not fully developed.

3.3.3 Time-averaged results

In Figure 3.14 the long-time averages of the vertical liquid velocity obtained with and without the BIT model of Sato are quantitatively compared to the experiment. The experimental data were taken at the centreline of the horizontal plane at different heights, 13cm, 25cm, and 37cm, as indicated in Figure 3.11. Both velocity profiles, with and without Sato model, shown in Figure 3.14 are almost the same at all three heights. Satisfactory agreement with the experiment is obtained at all heights except in the near wall region. It was observed that in the bubble column the time-averaged liquid flow is upward in the centre and downward close to the wall. The vertical liquid velocity changes its sign about 0.05 m away from wall with a slight trend that the position is closer to wall with increasing height. Such a trend for a broader distribution with increasing height is caused by a more homogenous gas distribution in the simulations with the increasing height. The similar results in both simulations can be attributed to the fact that in both cases the time-averaged eddy viscosity ratio $\langle\mu_{sgs}\rangle/\mu_{water}$ in the entire domain is not high enough for a significant change in the actual case. As shown in

Figure 3.15(c), the eddy viscosity without Sato model at the height 25 cm corresponds to about 5 times of the viscosity of the water. The Sato BIT model induced a slightly increased eddy viscosity in the centre, caused by the concentration of bubbles in this region (Figure 3.15b). But on the other hand, the absolute increase of the eddy viscosity is not very large due to the low gas superficial velocity as discussed by Ma et al. (2015b), so that the bubble plume dynamics are not damped substantially. All in all, the influence of the Sato BIT model on the profiles of the vertical liquid velocity is negligible at all three heights in Figure 3.14.

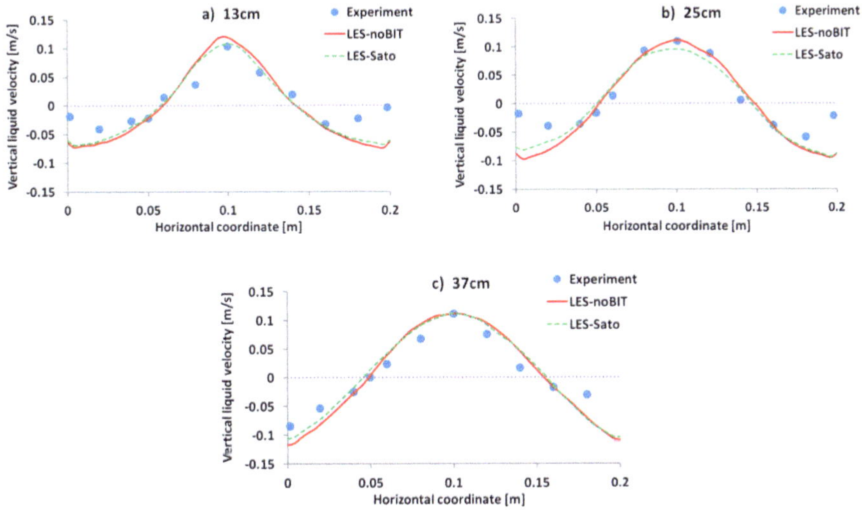

Figure 3.14: Comparison of time-averaged vertical liquid velocity with experimental data at various heights with and without Sato model, a) y = 13 cm, b) y = 25 cm, c) y = 37 cm.

Figure 3.15(a) presents a comparison of liquid Reynolds normal stresses and shear stress at the height 25 cm. Here, only the resolved part is shown using the difference between the instantaneous resolved fluid velocity and the time average of the resolved fluid velocity. The unresolved part is neglected because of the small SGS viscosity (Figure 3.15c). The Sato model has hardly any influence on the Reynolds stresses (Figure 3.15a). In the present conditions, the maximal $\mu_{bub}/\mu_{eff} \approx 25\%$ only, since the void fraction entering in (2.19) is small ($\alpha \leq 1.2\%$).

Compared to the vertical direction, the horizontal normal stress is smaller except in the centre region. Clearly, the velocity fluctuation in such a bubble column is anisotropic. From Figure 3.15(a), the peak of the time-averaged horizontal normal stress $\langle u'u' \rangle$ is located at the centre of the column, whereas the vertical fluctuations $\langle v'v' \rangle$ peak closer to the wall. This is in agreement with the findings of the similar bubble column experiments from Mudde et al. (1997) and Deen et al. (2000). And, indeed, the nature can be understood from the swinging motion of the bubble plume. In the centre, the horizontal liquid velocity u attains its highest magnitude, because the large vortical structures drive the bubble plume to span about half of the column width, while the rather uniform downflow or upflow near the wall does not contribute so much to the horizontal fluctuations. This leads the horizontal normal stress $\langle u'u' \rangle$ to peak in the column center. For vertical normal stress $\langle v'v' \rangle$, the phenomenon is the opposite. Close to the wall, the flow dynamically changes from upward to downward, depending on the location of the bubble plume. Therefore, the large fluctuations in the vertical component dominate in the near wall region rather than the centre region of the column, where the liquid moves almost primarily upwards. Both normal stresses are much higher than the Reynolds shear stress $\langle u'v' \rangle$. This was also reported by Mudde et al. (1997) from particle image velocimetry (PIV) data and by Akbar et al. (2012) from LDV data in two other rectangular bubble columns. This can be attributed to the fact that, in such quasi-2D bubble columns (i.e. columns with small deep), the vortical structures are regular and the maximal values of the liquid velocity fluctuation in vertical and horizontal direction do not simultaneously appear at the same time in the same position.

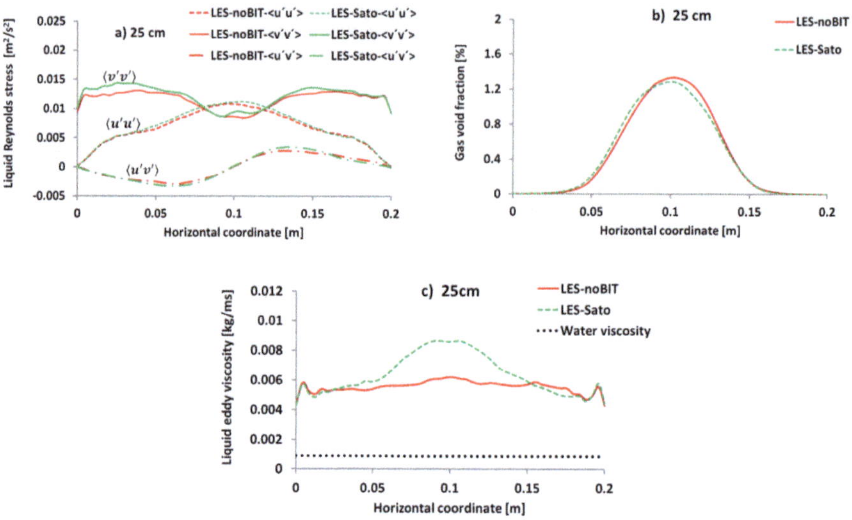

Figure 3.15: Profiles of statistical data at y = 25cm with and without Sato model. a) Reynolds stresses. b) Gas void fraction. c) Liquid eddy viscosity, compared to the molecular fluid viscosity.

Effect of numerical resolution

The governing equations for EE-LES are derived from applying a low-pass filter to the conservation equations of the two-fluid approach. In this study, as in most LES, the filter is not explicitly defined. It is assumed that discrete operations performed on a particular grid act as an implicit filter. Furthermore, the SGS model yields damping of the resolved fluctuations, related to the step size of the grid. Both issues cause the results to be sensitive to the mesh resolution (Fröhlich (2006), Bose et al. (2010)). To address this critical issue for the present case, an intensive mesh resolution study was performed without using any BIT model. Three different meshes were used as listed in Table 3.3. The coarse mesh consists of $20 \times 45 \times 8$ cells, resulting in a total number of only 7200 cells. The middle mesh was obtained by halving the grid size in each direction, resulting in 57600 cells. The finest mesh has totally 230400 cells with the filter width $\Delta = 2.7\ mm$. An even finer mesh is not needed here, because the time-averaged eddy viscosity shown in Figure 3.16(f) at the height 25 cm corresponds to twice the molecular viscosity of water only and hence the simulation is relatively close to the exact solution of two-fluid equations.

Mesh	I (coarse)	II (middle)	III (fine)
$\Delta x\ (mm)$	10	5	2.5
$\Delta y\ (mm)$	10	5	2.5
$\Delta z\ (mm)$	6.25	3.125	3.125
$\Delta = \sqrt[3]{\Delta x \Delta y \Delta z}\ (mm)$	8.55	4.275	2.7

Table 3.3: The three different meshes investigated.

Comparing the values for the vertical liquid velocity the middle mesh yields better agreement with the experiment than the coarse mesh at all the heights. The fine mesh provides further improvement close to the wall. A slight trend for a broader distribution with increasing height can be observed for all simulations. This is caused by a more homogenous gas distribution with increasing height in the simulations.

For the normal stress in both vertical and horizontal direction at height 25 cm, a somewhat too flat profile shown in Figure 3.16(e) is obtained with the coarse mesh. On the middle mesh and fine mesh the tendency of $\langle u'u' \rangle$ to develop a center peak and of $\langle v'v' \rangle$ to develop peaks near the wall gets more pronounced, as visible in Figure 3.16(d) and Figure 3.16(e), respectively. Since results of both, the liquid mean flow and the velocity fluctuations shown in Figure 3.16 are similar on the middle mesh and the fine mesh. It was decided to employ the middle mesh for cost reasons in further calculations.

In Figure 3.16(f), the eddy viscosity used in the 3 different meshes is represented. As expected from (3.4), the eddy viscosity with the coarse grid has the largest value, by a factor of 4 compared to the middle mesh and the eddy viscosity computed with middle mesh is about twice that with the fine mesh.

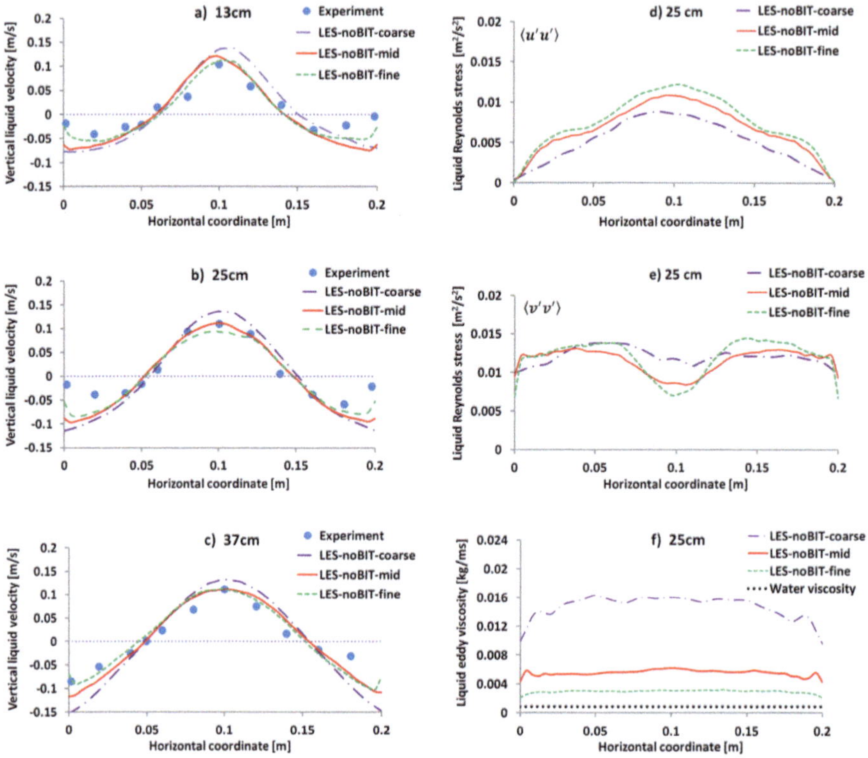

Figure 3.16: Comparison of time-averaged data from the mesh study conducted. The left three graphs show the vertical liquid velocity compared at three different heights with experimental data, a) y = 13 cm, b) y = 25 cm, c) y = 37 cm. The right three graphs show the liquid Reynolds normal stresses and the liquid eddy viscosities at y = 25 cm, d) vertical Reynolds stress, e) horizontal Reynolds stress, f) liquid eddy viscosity.

3.3.4 Analysis of spectra

The influence of the turbulence modelling can also be investigated by the transient behaviour at a fixed position. The time history of the horizontal liquid velocity is well suited for this purpose because this component clearly shows the oscillation of the bubble plume. Figure 3.17 shows experimental data, and different LES at the centre of the measurement line at the height of 25 cm (see Figure 3.11). The experimental data cover a time interval of 80 s (Pfleger, et al., 1999). For each simulation, 200 s from the beginning are presented in these figures corresponding to 10000 sample points. The periodic movement of the bubble plume is well reflected in the experimental graph. The superimposed high-frequency oscillations of the velocity are due to the turbulent fluctuations. The experimental time signal was not available electronically. To obtain an estimate of the low frequency contribution, the minimum being an estimate of the period, Figure 3.17(a) was sampled by hand eliminating in this procedure the high frequencies. The result of this operation is shown in Figure 3.18. The period in time determined from this smooth signal is about 12 s. The velocity amplitudes are in the range ±0.2 m/s. The signal between 125 s to 165 s in the experiment shows a bias towards positive values. This could reflect that the bubble plume is located only on one side of the column during this lap of time. Unfortunately a longer experimental time signal is not available. Both time signals from LES exhibit a similar transient behaviour as the experimental signal in terms of the amplitude of the fluctuations. The results obtained with the dynamic model without BIT model show a large scale period of about 12 s, which is fairly constant in time. In the other simulation employing the BIT model the fundamental period is not that constant and a tendency towards intermittency is noticed.

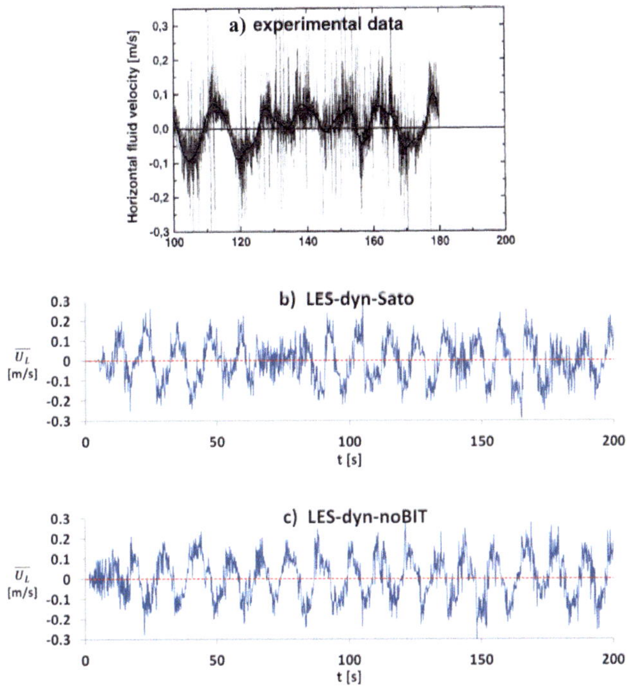

Figure 3.17. Time history of the horizontal liquid velocity at the center of the measurement line at the height of 25 cm. a) Data obtained from the experiment (Pfleger, et al., 1999). [Graph reproduced with permission]. b) Time signal from the simulation with BIT model. c) Time signal from the simulation without BIT model.

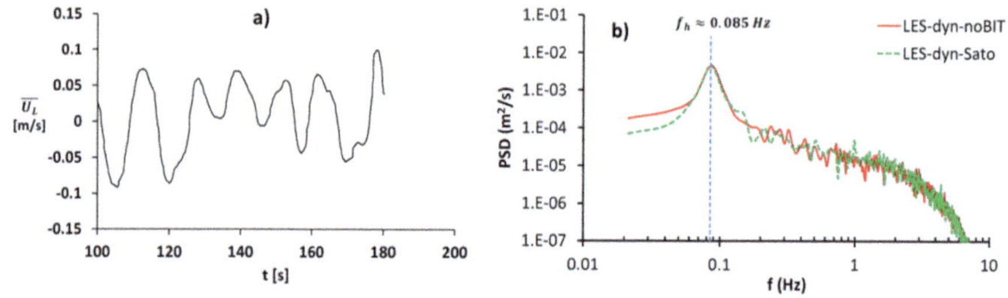

Figure 3.18: Analysis of 1-D time signals at the measurement line at the height of 25 cm above the bottom in the center of the container. a) Estimation of the low-pass filtered horizontal liquid velocity described in the text. b) Power spectrum densities of the horizontal liquid velocity obtained in the two simulations with and without BIT model.

The differences observed in the simulations are caused by the BIT contribution, so that an investigation of the spectra is helpful. The power spectrum densities (PSD) obtained from LES with and without BIT model are shown in Figure 3.18(b). The data were extracted from the centre of the measurement line and cover the time from 50 s to 500 s. The principle oscillation in the experiment (Figure 3.18a) is well reproduced by both simulations, as shown in the PSD diagram, with the same magnitude peaking at about 0.085 Hz. The slopes in the dissipation range of the spectrum with and without Sato BIT model seem to be the same, although the one with the Sato BIT model has a little higher effective viscosity, especially in the centre of the column (Figure 3.15c). Normally, with a higher effective viscosity, the spectrum is more damped (Fröhlich, 2006). In the present case, however, the behaviour is different, because the averaged increase of the eddy viscosity (in the column centre is about 33%) has a similar effect as increasing the Smagorinsky constant C_s from 1.0 to 1.15 in an LES using the Smagorinsky model. Such a small difference could not be seen in the statistical evaluation in a LES for a similar bubble column (Sungkorn et al. (2011) and Zhang et al. (2006)).

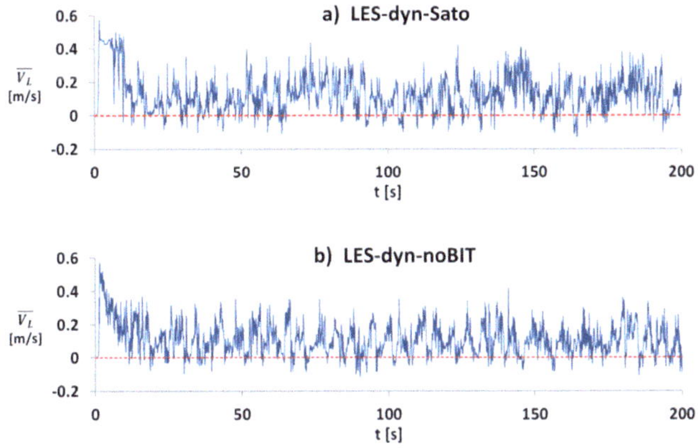

Figure 3.19: Time history of the vertical liquid velocity at the center of measurement line at the height of 25 cm. a) Time signal from the simulation with BIT model. b) Time signal from the simulation without BIT model.

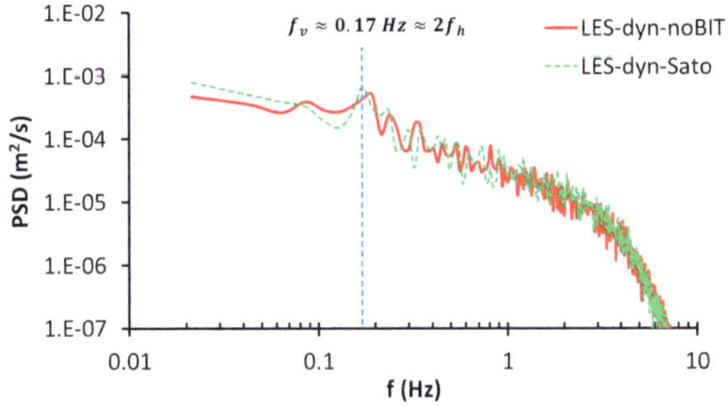

Figure 3.20: Power spectrum density of vertical liquid velocity obtained with and without BIT model at the center of measurement line at the height of 25cm.

The vertical velocity component at the measurement point is now addressed as well (experimental data for this component are not available). Two 200 s time history plots of the resolved of the vertical liquid velocity at this position are reported in Figure 3.19. The magnitude of the vertical velocity is similar in both simulations, varying roughly between 0 m/s and 0.3 m/s. The corresponding PSD diagrams are shown in Figure 3.20. Both are very similar with the same magnitude peaking at about 0.17 Hz. It is in line with the understanding of the flow that the main frequency in vertical direction is about twice the horizontal one, since the movement of the plume introduces a change from positive over negative to positive vertical fluctuation by moving once from left to right, i.e. going once through maximum or minimum. The liquid vertical velocity in the column centre is almost permanently upwards as shown in Figure 3.19. Whether the bubble plume is on the left side of the column centre or on the right side cannot be deduced from the data in Figure 3.19.

4 Scale-adaptive simulation of a square bubble column

In the present chapter, a second type of scale-resolving simulation, here, the so-called scale adaptive simulation (SAS) is performed for liquid phase in a square cross-sectioned bubble column. SAS aims to resolve the unsteady large-scale turbulent motion without explicit grid dependence and was for this reason, termed second generation URANS models (2G-URANS) (Fröhlich & von Terzi, 2008). The turbulence in the dispersed gas phase is modelled with the same zero equation as for the previous chapter. The results are compared to the experimental data from Deen et al. (2000) and to own LES. Furthermore, the effect of modelling bubble-induced turbulence with SAS and LES is discussed. The results of this chapter have been published in Ma et al. (2015b).

4.1 Description of method employed

SAS for continuous liquid phase

The unresolved stress tensor $\tau_{ij} = \overline{u_i u_j} - \bar{u}_i \bar{u}_j$, is modelled by the SAS model of Menter & Egorov (2010). It employs an equation for the turbulent kinetic energy k and an equation for $\Phi = \sqrt{K}\, L_t = \upsilon_t c_\mu^{-1/4}$, with L_t the turbulent length scale, and the components of $\boldsymbol{u}$ denoted u_j, reading:

$$\frac{\partial(\rho_L k)}{\partial t} + \frac{\partial(\rho_L \bar{u}_j k)}{\partial x_j} = P_K - c_\mu^{3/4} \rho_L \frac{k}{\Phi^2} + \frac{\partial}{\partial x_j}\left(\frac{\mu_t^L}{\sigma_K}\frac{\partial k}{\partial x_j}\right), \tag{4.1}$$

$$\frac{\partial(\rho_L \Phi)}{\partial t} + \frac{\partial(\rho_L \bar{u}_j \Phi)}{\partial x_j} = \frac{\Phi}{k} P_k\left(\zeta_1 - \zeta_2\left(\frac{L_t}{L_{vk}}\right)^2\right) - \zeta_3 \rho_L k + \frac{\partial}{\partial x_j}\left(\frac{\mu_t^L}{\sigma_\Phi}\frac{\partial \Phi}{\partial x_j}\right). \tag{4.2}$$

Here, L_{vk} is the von Karman length-scale defined by

$$L_{vk} = \kappa \frac{S}{|\bar{u}''|}, \quad S = \sqrt{2 S_{ij} S_{ij}},$$
$$S_{ij} = \frac{1}{2}\left(\frac{\partial \bar{u}_i}{\partial x_j} + \frac{\partial \bar{u}_j}{\partial x_i}\right), \qquad |\bar{u}''| = \sqrt{\frac{\partial^2 \bar{u}_i}{\partial x_j \partial x_j}\frac{\partial^2 \bar{u}_i}{\partial x_k \partial x_k}}, \tag{4.3}$$

and

$$P_K = \mu_t S^2. \tag{4.4}$$

The values of the constants are $\zeta_1 = 0.8$, $\zeta_2 = 1.47$, $\zeta_3 = 0.0288$, $\sigma_\emptyset = 2/3$, $\kappa = 0.41$, $c_\mu = 0.09$, according to (Menter & Egorov, 2010). The information provided by the von Karman length-scale allows the SAS model to treat large turbulent structures in a URANS fashion so that less computational effort is required compared to LES. The L_{vk} term can be transformed and implemented into any other scale-defining equation. Here, the Shear Stress Transport model (SST) is used, resulting in an SAS-SST model. Menter & Egorov (2005) introduced an additional SAS source term F_{SAS} in the ω-equation resulting from the transformation. The finally used version of this source term reads (Egorov & Menter, 2008):

$$F_{SAS} = max\left\{\rho_L \zeta_2 \kappa S^2 \left(\frac{L}{L_{vK}}\right)^2 - C\frac{2\rho_L k}{\sigma_\Phi} max\left\{\frac{1}{\omega^2}\frac{\partial\omega}{\partial x_j}\frac{\partial\omega}{\partial x_j}; \frac{1}{k^2}\frac{\partial k}{\partial x_j}\frac{\partial k}{\partial x_j}\right\}; 0\right\}, \tag{4.5}$$

with $\zeta_2 = 3.51$ and $C = 2$. The equilibrium eddy viscosity of the SAS-SST model is:

$$\mu_t^L = \rho^L \left(\sqrt{\left((\beta/C_\mu) - \alpha\right)/(\kappa\zeta_2)} \cdot L_{vk}\right)^2 S. \tag{4.6}$$

with $\beta = 0.075$. This formula has a similar structure as the Smagorinsky SGS eddy viscosity used for the LES in this study as described below. Then, the SAS model is employed in its two-phase version, with the void fraction α^L incorporated in a straightforward way into each term of (4.1) and (4.2). Furthermore, for SAS, up to now, non-BIT models are available in the literature. For these reasons, the common BIT model of Sato et al. (1981) is employed here and represented by an additional extra term in the SGS turbulent viscosity as described in (3.5).

Because the liquid is stagnant at the beginning, the initial condition for the turbulence parameters are considered only for the numerical stability, defined as: liquid turbulent kinetic energy $k = 3.75 \cdot 10^{-7}\, m^2/s^2$ and eddy frequency $\omega = 0.042\, s^{-1}$ (Ansys, 2013).

Additionally, results obtained with EE-LES for the same configuration and with the same computational grid are used for comparison reason. They were obtained with the Smagorinsky SGS model based on the eddy viscosity (3.3). The model constant C_s was chosen to be $C_s = 0.12$ and full details of this approach for two-phase flow are provided in Section 3.1.

Interfacial forces

The corresponding interfacial transfer models employed in the present chapter are the same as in EE-LES in the previous chapter except the different usage of the turbulent dispersion force model for different turbulence approaches. Here, the turbulent dispersion model of Burns et al. (2004) is not employed with LES, but in SAS. Since using SAS the unresolved part is often not negligible, and larger than in LES.

4.2 Simulation of the bubble plume

4.2.1 Experimental data and simulation setup

The simulations were carried out for an air/water bubble column at ambient pressure (Deen, et al., 2000), schematically shown in Figure 4.1. The bubble column has a square cross-section of 0.15 m × 0.15 m and is filled with distilled water up to a height of 0.45 m. A distributor plate containing 49 needles with an inner diameter of 1 mm was placed in the centre of the bottom with a square section of $A_{in} = 0.03\ m \times 0.03\ m$. Measurements were performed for the gas superficial velocity of 4.9 mm/s and took place 0.25 m above the distributor plate (y = 0.25 m) in the centre plane (z = 0.075 m). A particle image velocimetry system was used to simultaneously measure the gas and liquid velocities. In order to obtain a non-coalescing system, 4 g of kitchen salt per litre water was used. The bubble size was 4 mm (Deen, et al., 2001). More details on the experiment are provided in that reference.

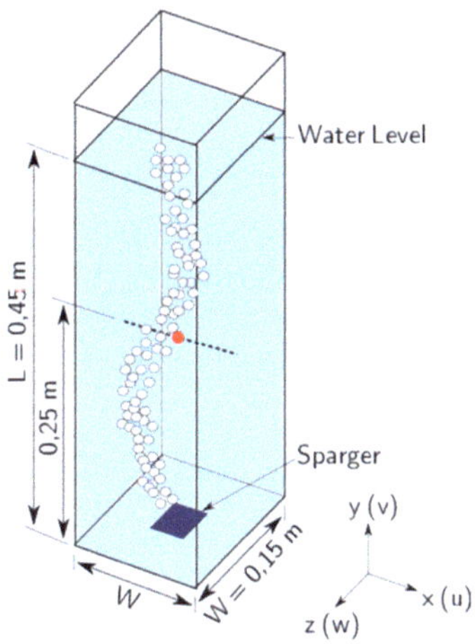

Figure 4.1: Geometry of the bubble column experiment of Deen et al. (2000). The broken line shows the measurement position. The marker on that line is the measurement point for the results presented in Figure 4.10, Figure 4.11, and Figure 4.12.

The gas inlet is defined as a distributed source in the central area $0.03 \times 0.03\ m^2$ at the bottom of the domain, representing the experimental needle setup. A superficial gas velocity of 4.9 mm/s is imposed leading to a gas velocity u_G at the inlet of $0.12\ m/s$ with $\alpha^G = 1$ and $\alpha^L = 0$. The other boundary and initial conditions are the same as for the cases in Section 3. For the spatial discretization, a central difference scheme is employed, and a second-order backward Euler scheme is used in time. The simulations were carried out using a time step of $\Delta t = 0.0025 \cdots 0.005\ s$ to satisfy CFL < 1. The simulations were run for 200 s.

The rectangular bubble column was discretized with uniform cubic cells of $\Delta x = \Delta y = \Delta z = 5\,mm$, resulting in 81,000 cells, overall. The bubbles are treated as mono-disperse with bubble diameter $d_p = 4\,mm$, according to the experiment. In the present work, the container is relatively small compared to the bubble size: $w = 150\,mm$ and $d_p \approx 4\,mm$ yields a factor of $w/d_p \approx 38$. Discretizing the container cross section with 30×30 cells allows reliable discretization of derivatives, while substantially coarser grids do not. This grid is well suitable for the Euler-Euler method and hence selected here according to the mesh studies of Niceno et al. (2008) and Sungkorn et al. (2011) to supply the best prediction. The simulations were conducted with the software Ansys CFX 14.5 using a finite volume solver.

4.2.2 Transient results

Snapshots of isosurfaces of constant void fraction with SAS and LES are shown in Figure 4.2 together with the corresponding contour plots of the void fraction in the centre plane. On average, the bubbles reach the top of the column after about 1.2 s. The isosurfaces provide a good illustration of the dynamics of the bubble plume and look very similar to the numerical results presented in Niceno et al. (2009). For LES, the plume is more distorted than in case of SAS. This is caused by the smaller effective viscosity compared to SAS, leading to a better resolution of small fluctuations. The corresponding void fraction in the centre plane shows a difference between the results for the two turbulence models. In the results of SAS, due to the application of turbulent dispersion model, the gas volume fraction is distributed somewhat more homogenously over the cross section. It was confirmed by an own simulation (not reported here), that using SAS without turbulent dispersion model yields a bubble plume rising up to the top of the column without much transversal spreading. This is in contrast to the experiment, where the bubble plume moves around in the bubble column in a random fashion (Deen, et al., 2001).

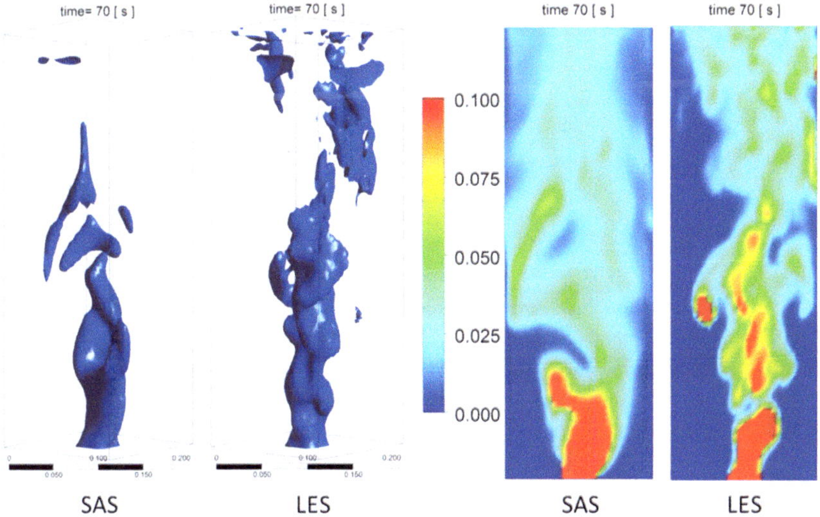

Figure 4.2: Instantaneous data from SAS and LES at $t = 70\ s$. The left two graphs show instantaneous isosurfaces of constant void fraction ($\alpha = 0.05$). The right two graphs show the instantaneous void fraction in the centre plane at the same instant.

4.2.3 Time-averaged results

In this section, the simulation results averaged from $t = 20\ s$ to $t = 200\ s$ are quantitatively compared with the experiment. The experimental data were taken at the centreline of the horizontal plane at height 25 cm, as indicated in Figure 4.1. Time averaged vertical liquid and gas velocity, vertical and lateral liquid velocity fluctuation, gas void fraction and effective viscosity are presented for both cases.

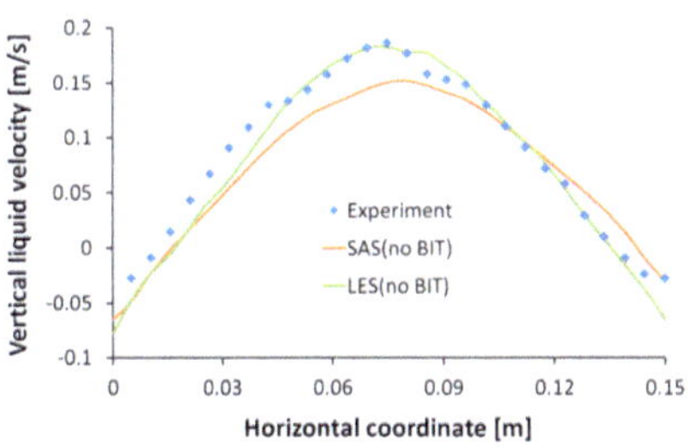

Figure 4.3: Vertical component of liquid velocity obtained from SAS, LES without BIT model and experiments.

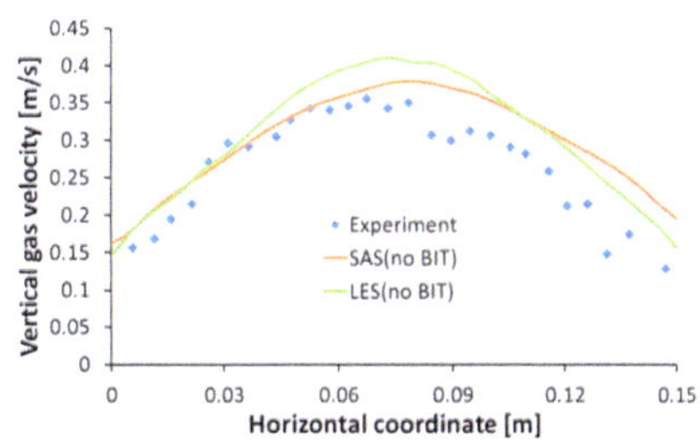

Figure 4.4: Vertical component of gas velocity obtained from SAS, LES without BIT model and experiments.

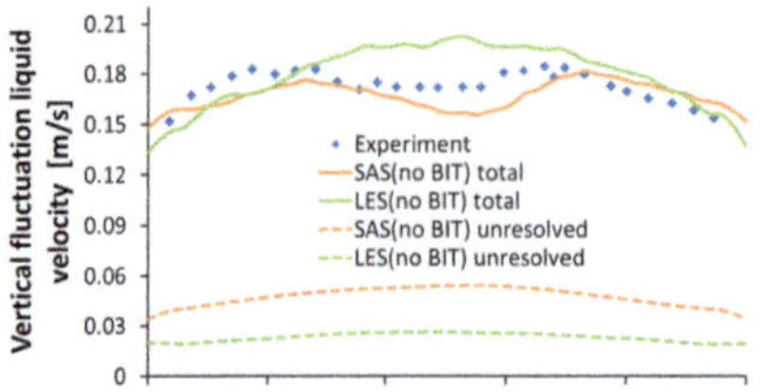

Figure 4.5: Fluctuations of vertical liquid velocity component obtained from SAS, LES without BIT model and experiments. Continuous lines indicate the total fluctuations obtained from adding the resolved and the unresolved part. The unresolved part computed as explained in the text.

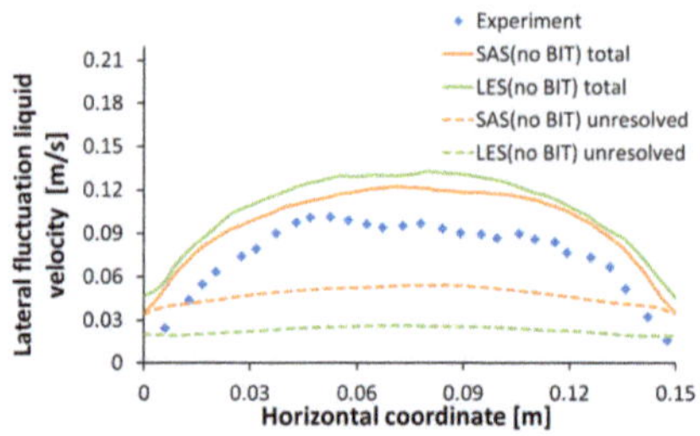

Figure 4.6: Fluctuations of lateral liquid velocity component obtained from SAS, LES without BIT model and experiments. Continuous lines indicate the total fluctuations obtained from adding the resolved and the unresolved part. The unresolved part computed as explained in the text.

Figure 4.3 and Figure 4.4 show profiles of the averaged vertical liquid and gas velocity. Satisfying quantitative agreement with the experiment is obtained for both simulations. The results of SAS have a bit lower values compared to the LES in both profiles. The liquid velocity profile is somewhat underpredicted with SAS, but well reproduced using LES. On the other hand, the gas velocity agrees better using SAS compared to the LES. This effect can be attributed to the drag force modelling and should be subject of further investigations. In both simulations the relative velocity between liquid and gas is about 0.2 m/s, i.e. it is nearly the same, resulting from the use of the same drag force model. The vertical liquid velocity changes its sign about 0.02 m away from wall, which is in agreement with the experimental data. The downflow in the near-wall region is caused by the liquid mass balance in the bubble column.

Figure 4.5 and Figure 4.6 present a comparison of the vertical and lateral fluctuating liquid velocity. In the previous studies by Deen et al. (2001) and Dhorte et al. (2008), only the resolved velocity fluctuations were provided, while the unresolved part was neglected. In the present SAS and LES, the total velocity fluctuation is considered, summing the resolved part $\bar{u}''$ and the unresolved part u' as mentioned in Section 3.1. For the resolved part, $\bar{u}''\bar{u}'' = [\bar{u} - \langle\bar{u}\rangle]^2$ is computed using the difference between the instantaneous resolved velocity $\bar{u}$ and the time average of the resolved velocity, $\langle\bar{u}\rangle$. Modelling the unresolved part is different for SAS and LES. SAS has a transport equation for the turbulent kinetic energy. Assuming isotropy of turbulence leads to $\overline{w'w'} = \overline{v'v'} = \overline{u'u'} = \frac{2}{3}k$. For LES, only the anisotropic part of the SGS stress tensor is considered in the Smagorinsky model, so that the information about the isotropic part, $\tau_{kk} = 2k$, is lost. Here, the estimation with equation (3.6) for k_{sgs} is used (Menter, 2013).

The qualitative comparison of the fluctuations of the vertical liquid velocity shown in Figure 4.5 is encouraging for both, SAS and LES. However, the twin-peaked shape seen in the experiments is only reproduced by the SAS. The unresolved vertical velocity fluctuation with the SAS is higher than that with the LES, due to the higher effective viscosity of the SAS (see Figure 4.8 below). This effect is expected and matches the basic concept of corresponding hybrid LES/RANS methods (here SAS) as discussed by Fröhlich & von Terzi (2008). The fluctuating lateral liquid velocity shown in Figure 4.6 is not predicted as well as the one of the vertical component. Quantitatively, both SAS and LES overpredict the magnitude to some extent. This overprediction is caused by the high ratio of the modelled lateral fluctuations to the total lateral fluctuations (about 35% in SAS and 16% in LES). Compared to the vertical direction, the resolved fluctuations in lateral direction are much smaller. But determining the unresolved parts isotropic turbulence is assumed with both turbulence models and provides the same values as for the vertical direction.

Effect of using the Sato Model for BIT

Previous studies of Deen et al. (2001), Dhorte et al. (2008) and Niceno et al. (2008) mentioned that there was no significant change in the predictions using LES with and without the BIT model of Sato. This is not confirmed in the present work.

Figure 4.7 shows the mean vertical liquid velocity profiles with and without the Sato model for both turbulence models. There is a more pronounced peak in the centre with both approaches, when the model is included, clearly overpredicting the experimental data. This can be attributed to the fact that, as shown in Figure 4.8, the additional viscosity caused by the Sato BIT model in (2.19) yields an increased effective viscosity in both SAS and LES, which in turn damps the bubble plume dynamics. A behaviour similar to Figure 4.7, with an even higher centre peak in the vertical liquid velocity profile was obtained by Zhang et al. (2006). He performed LES with an increased value of $C_S = 0.2$ in the Smagorinsky SGS model instead of $C_S = 0.12$. The application of the Sato model in the LES has a similar effect as increasing C_S, since it enhances the SGS viscosity.

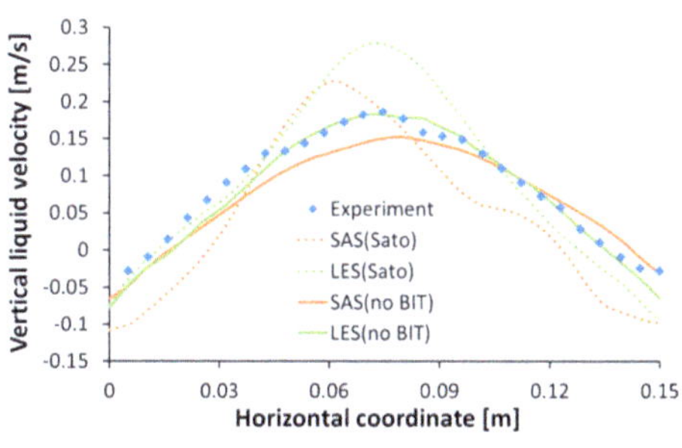

Figure 4.7: Vertical liquid velocity obtained from SAS, LES and experiments, with and without Sato model.

Figure 4.9 presents the comparison of the void fraction with and without the Sato model. Using the model in both SAS and LES, a more pronounced peak in the centre is obtained, compared to a flatter profile without the model. The larger gas volume fraction in this region also increases the liquid velocity in the centre and thus is a second reason for the steeper

velocity profiles. Unfortunately, a comparison with the experimental data is not possible, because no void fraction profiles were measured.
In the earlier results of Deen et al. (2001), Dhorte et al. (2008) and Niceno et al. (2008) who also used the Sato model with LES, an increase of the vertical velocity in the centre was not observed as in the present work. One reason could be the fact that these authors used a different value of the lift coefficient, $C_l = 0.5$. In the present study, the lift force correlation of Tomiyama is used with a lower effective lift coefficient $C_l \approx 0.28$, along with a turbulent dispersion force that also results in dispersing the bubble plume. As reported by Lucas & Tomiyama (2011), the lift force may have a significant effect on the gas plume transverse spreading.

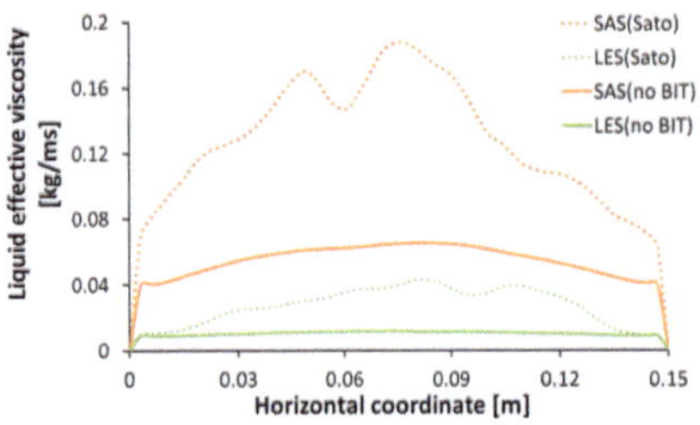

Figure 4.8: Liquid phase effective viscosity obtained from SAS and LES, with and without Sato model.

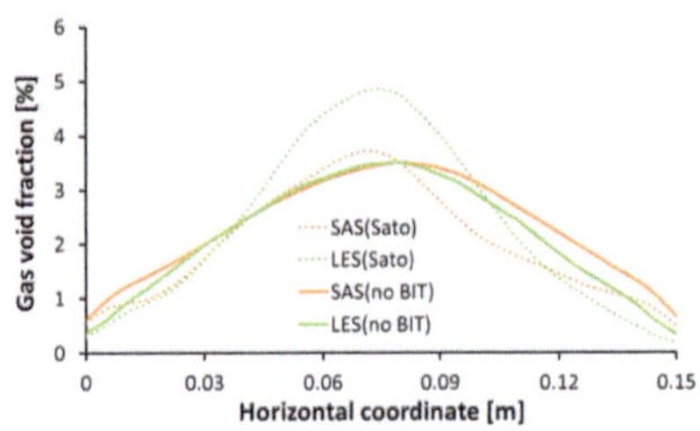

Figure 4.9: Gas void fraction obtained from SAS, LES, with and without Sato model.

4.2.4 Analysis of spectra

Figure 4.10 and Figure 4.11 show time series of the resolved vertical liquid velocity at the centre of the measurement line shown in Figure 4.1 for SAS and LES without BIT model. The period of 50 s presented in these figures corresponds to 189 sample points for the experimental data and 500 sample points for the simulations. Both, SAS and LES reflect the transient behaviour observed in the experiment of Deen et al. (2001) in terms of the amplitude of the fluctuations. The differences observed in the signals are caused by the different sampling rates, so that an investigation of the spectra is helpful.

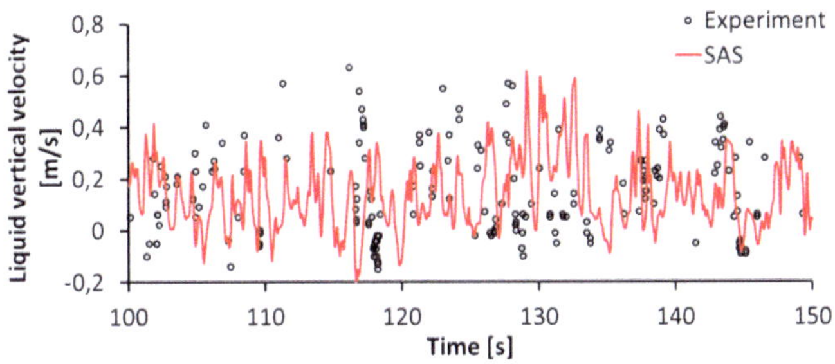

Figure 4.10: Time history of the liquid vertical velocity obtained with SAS at the centre of measurement line, compared to experimental data from Deen et al. (2001).

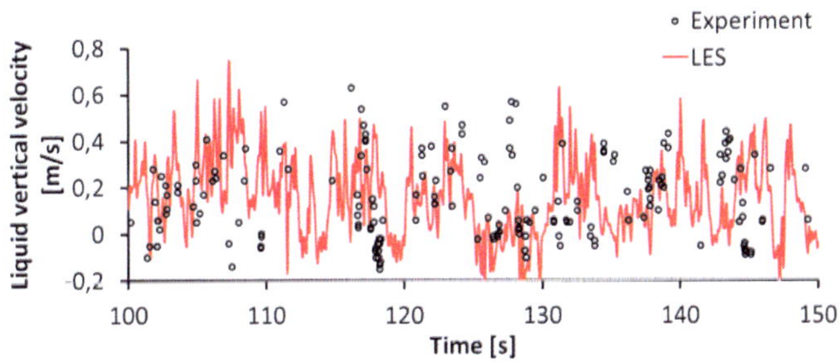

Figure 4.11: Time history of the liquid vertical velocity obtained with LES at the centre of measurement line, compared to experimental data from Deen et al. (2001).

The PSD of SAS and LES without BIT model obtained with 18000 sample points are shown in Figure 4.12. The data were also extracted from the centre of the measurement line and cover the time interval $t = 20\ s$ to $t = 200\ s$. A fast Fourier transform of the vertical liquid velocity signal using the Welch method (Welch, 1967) with non-overlapping sections and a Hanning window was performed. Because of the low sample frequency, the experimental spectrum can only be determined up to about 4 Hz.

Both spectra with SAS and LES exhibit a broad range of frequencies and slopes of about -5/3 in the inertial subrange, like single phase turbulent flow. Both predictions are in good agreement with the experiment in the low frequency region. Hence the large-scale vortical magnitudes are well reflected in the simulations. Compared to LES, the spectrum of SAS is more damped caused by the larger eddy viscosity. This is a similar effect as increasing the Smagorinsky constant C_s in a LES, which is discussed by Fröhlich (2006).

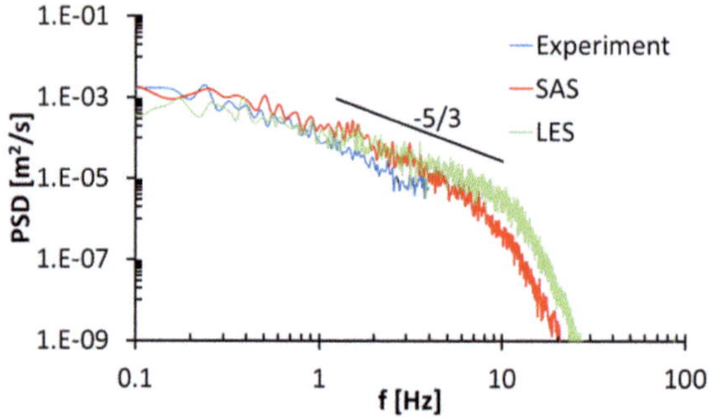

Figure 4.12: Power spectrum density of vertical liquid velocity taken at the centre of measurement line for SAS and LES in comparison to the result obtained from the experimental time signals.

5 Improved RANS model based on DNS data

As mentioned in the introduction, in single-phase flow DNS data have been used as a basis for development of improved RANS models and have resulted in significant progress. The present chapter will follow this strategy for bubbly flow.

The starting point of the present work is the equations for two-phase flows rigorously derived by Kataoka & Serizawa (1989). They are based on a single-phase representation including the influence of the bubbles by additional interfacial terms in the balance equations, both for k, the TKE, and ε, the dissipation rate of the TKE. When supplementing the unclosed terms in these equations with adequate models they constitute an appropriate framework for RANS modelling of bubbly flows. Due to the lack of reliable data for individual terms, however, the form of these models is mostly based on *ad hoc* physical considerations, e.g. by Troshko & Hassan (2001), Politano et al. (2003), and Colombo & Fairweather (2015).

Recently, direct numerical simulation (DNS) of sufficiently large scale turbulent bubbly flows have become available to shed some light on the very details of the physics involved (Roghair et al. (2011a); Lu & Tryggvason (2013)). Ilić (2006) performed DNS studies of monodisperse bubbles rising in initially stagnant liquid within a plane channel and evaluated each term in the TKE budget for up to eight bubbles. She demonstrated for that special case that the gain of TKE is mainly caused by the interfacial term while the contribution of the production term is negligible. Besides evaluating the relative importance of each term in the TKE budget under corresponding conditions, the DNS data were also used for *a priori* test of available BIT models to assess the accuracy of them. Erdogan & Wörner (2014) extended this work to larger bubble Reynolds numbers, Re_p up to 230, against maximal 56 in Ilić (2006), showing that the TKE of the liquid phase increases with Re_p.

To the best of the authors' knowledge, DNS data have not been used so far to generate a complete Euler-Euler RANS model that can be applied to bubbly flows. Here, such an approach is used employing the data of Santarelli & Fröhlich (2015a) (2016b) to devise a new model for the BIT term that dominates in many disperse bubbly flows. Two main issues are investigated, the length and time scale of the BIT source in the dissipation equation and the coefficients in the BIT source expressions of the two-equation RANS model used.

The remainder of the chapter is organized in seven sections. In Section 5.1, the DNS data severed for BIT modelling are introduced. Energy spectra were made to identify the suitable length and time scale for the source term in the dissipation equation in Section 5.2. The BIT modelling within framework of EE-RANS and the model parameters pertaining to the simulations reported later are described in Section 5.3 and Section 5.4, respectively. In Section 5.5, an improved BIT model is introduced and critical points in previous BIT approaches are discussed in Section 5.6. The new BIT model is then validated with an experiment in Section 5.7. Some results reported in this chapter have been presented by Ma et al. (2017).

5.1 DNS database

To serve the present purpose the DNS data have to provide information about all relevant processes, in particular the interfacial effects, requiring that the bubble geometry is resolved. This is in contrast to the point-particle approach, occasionally also addressed as DNS of a multiphase flow (Balachandar & Eaton, 2010). Next consideration is then given to choosing appropriate DNS data for this special issue, BIT modelling. Regarding to the some DNS data e.g. by Ilić (2006) and Erdogan & Wörner (2014), strictly speaking, such DNS data are insufficient to develop BIT models in the framework of EE-RANS. This is due to the small dimensions of their considered configurations, which results in a very low number of bubbles (maximal 8 bubbles in Ilić (2006) and 6 in Erdogan & Wörner (2014)). These cases are quite far from the continuum assumption in the two-fluid model, i.e. for bubble swarms with huge numbers of bubbles; the gaseous phase can be modelled as a continuum and represented by a void fraction. Besides considering the number of bubbles due to the two-fluid approach, special attention should also be taken in the strategy of the RANS modelling for an explicit term in the turbulence energy equation. Although the TKE is a scalar quantity, noting its response in various situations has proved very helpful in guiding modelled forms of equations (Hanjalić & Launder, 2011). For instance, as illustrated in Figure 5.1 from Hanjalić & Launder (2011), energy decay in homogenous turbulence is a typical case to determine the empirical coefficient $C_{\varepsilon 2}$ associated with the sink term in the ε-equation for single-phase flow; since there the dissipation term is dominant. In an analogous manner, here, the DNS cases dominated by BIT or more precisely, dominated by the interfacial term in the exact k-equation are searched for modelling the BIT source term. Of such interests described above, to the best of the authors' knowledge, may be the recent DNS data from the group of Jochen Fröhlich in Dresden, Germany are the only option. In Santarelli & Fröhlich (2015a) (2016b), bubble-resolving DNS with many thousands of bubbles at low Eötvös number were conducted by an Immersed Boundary Method (Uhlmann (2005), Kempe & Fröhlich (2012)). In these DNS cases bubbles were considered as spherical objects of fixed shape and a no-slip condition was applied at the phase boundary matching with the behaviour of air bubbles rising in the contaminated water. The technically involved numerical procedure to evaluate the TKE budget was presented in Santarelli et al. (2016a).

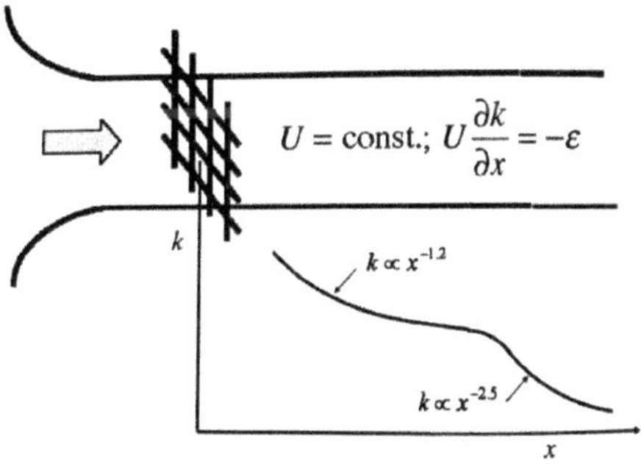

Figure 5.1: Schmeatic representation of the experiment for energy decay in homogeneous turbulence along the tunnel (Hanjalić & Launder, 2011). [Figure reproduced with permission from Cambridge University Press]

The DNS were conducted for upward flow between two flat vertical walls in a rectangular channel, with periodicity in streamwise (x) and spanwise (z) direction and on both walls a no-slip condition for liquid was applied. The size of the domain is $L_x \times L_y \times L_z = 4.41\mathrm{H} \times \mathrm{H} \times 2.21\mathrm{H}$, where H is the distance of the walls. Figure 5.2 shows the domain and an instantaneous snapshot of the bubbly flow in one of the DNS cases, labelled *SmMany*. The gravity force acts in the negative x-direction and the bulk velocity U_b was kept constant by instantaneously adjusting a volume force, equivalent to a pressure gradient, thus imposing the same bulk Reynolds number $Re_b = U_b H/\nu$, where ν is the kinematic viscosity of the liquid. The DNS data used in this work were obtained for three monodisperse cases (*SmMany, SmFew, LaMany*) and one bi-disperse case labelled *BiDisp*, of the same void fraction as *SmMany* and *LaMany* with half the void fraction consisting of small bubbles and the other half of larger bubbles, all at the same bulk Reynolds number $Re_b = 5263$. Additionally, a single-phase simulation labelled *Unladen* was performed under the same conditions for comparison. The domain was discretized for all DNS cases by a cubic mesh of $1024 \times 232 \times 512$ points in the streamwise, wall-normal and spanwise directions with the same step size $\Delta = H/232$. The bubble Reynolds number Re_p, based on the relative velocity between phases and the bubble diameter d_P, is between 200 and 500. Table 5.1 provides an overview of all cases. The data available cover statistical moments of first and second order for liquid and bubbles as well two-point correlations. More details are provided in the cited references.

Case	*Unladen*	*SmMany*	*SmFew*	*LaMany*	*BiDisp (Sm/La)*
N_p	-	2880	384	913	1440/456
α	-	2.14%	0.29%	2.14%	1.07%/1.07%
d_P/Δ	-	12.1	12.1	17.4	12.1/17.4
d_P/H	-	0.052	0.052	0.076	0.052/0.076
Re_p	-	235.5	268.3	475.2	233.6/463.6
C_D	-	0.89	0.705	0.666	0.93/0.703
Re_b	5263				
$L_x \times L_y \times L_z$	$4.41H \times H \times 2.21H$				
$N_x \times N_y \times N_z$	$1024 \times 232 \times 512$				

Table 5.1: Parameters of the DNS cases used for the present modelling. The labels *Sm* and *La* used for the case with the bi-disperse bubble swarm, *BiDisp,* indicate averaging only over the small and large bubbles, respectively. Here, N_p is the number of bubbles, α the void fraction, d_P the particle diameter. The values of Re_p, the particle Reynolds number, and C_D, the drag coefficient, are results of the DNS.

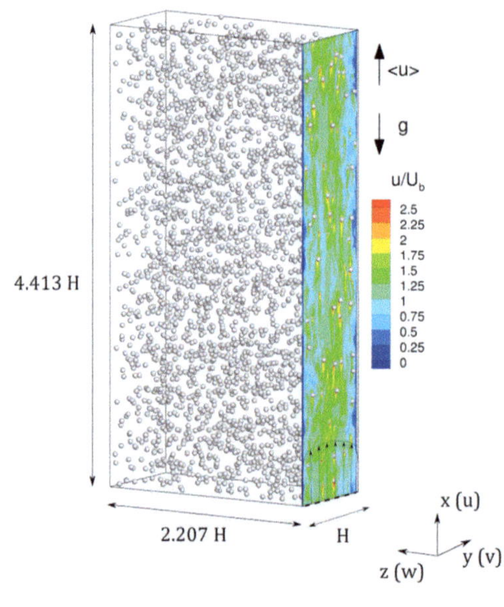

Figure 5.2: Instantaneous DNS data for the case *SmMany*, the bubble size is to scale and the vertical plane with the contour plot shows the instantaneous streamwise liquid velocity (Santarelli, 2015). [Figure reproduced with permission from TUD Press]

Redimensionization of DNS settings

As mentioned in the introduction, the RANS modelling for BIT in the present work embedded within the framework of two-fluid model, in which all momentum exchange between phases are modelled. The closure models for such momentum exchange need empirical correlations, which are almost directly related to the physical bubble size (see Section 2.2), with the corresponding simulations conducted in dimensional units (Ishii & Hibiki, 2011).
However, the DNS were performed using a non-dimensional set of parameters. In single-phase simulations, the flow between parallel plates is well characterized by the Reynolds number to describe the relationship between fluid system (specified by fluid velocity and fluid viscosity) and channel geometry (specified by channel width H). For each physical dimensional quantity of interest, *A*, following notation is used

$$A = \tilde{A}A^*. \tag{5.1}$$

Here, $\tilde{A}$ is a dimensional reference value of the quantity and the superscript, $*$, denotes a non-dimensional parameter used in the DNS. In the single-phase DNS of the *Unladen* case following non-dimensional parameters are used $\rho^* = 1, H^* = 1, U^* = 1, \nu^* = 0.00019$, for liquid density, channel width, mean liquid velocity, and liquid kinematic viscosity, respectively, which yields the bulk Reynolds number $Re_b = \frac{U^*H^*}{\nu^*} = 5263$. If we assume the liquid is water with the physical dimensional values at normal conditions: $\rho = 1000\ kg/m^3$, $\nu = 8.899 \cdot 10^{-7} m^2/s$, the reference values with $\tilde{\rho} = \frac{\rho}{\rho^*} = 1000\ kg/m^3$ and $\tilde{\nu} = \frac{\nu}{\nu^*} = 4.6837\ m^2/s$ are then fixed.
For two-phase flows, the situation is more complicated due to the physical representation of the bubble diameter d_P. Here, the bubble diameter is actually a separate parameter independent from the channel geometry and the fluid system, i.e. although there is a fixed scaling between bubble diameter and the channel width in DNS (e.g. $\frac{d_P}{H} = 0.052$ for the *SmMany* case, see Table 5.1), choosing an arbitrary dimensional length scale for the bubble diameter or the channel width is critical. One apparent example is that bubbles with $d_P = 1mm$ passing the channel of width $H = 10\ mm$ have totally different physics compared to the bubbles with $d_p = 10\ mm$ rising in a channel of width $H = 100\ mm$. Such an arbitrary choice of the length scale can lead to an incorrect representation of the real effect of buoyancy in the corresponding DNS and then, the non-dimensional relative velocity U_r^* calculated in DNS after redimensionization cannot fit the physical one.
To solve this problem, another dimensionless parameter has to be introduced to describe the relationship between bubble diameter and the fluid system for a more accurate interpretation of the bubbly flow in DNS. This is done by the Archimedes number, Ar, defined as

$$Ar = \frac{(\rho^L - \rho^G) g {d_P}^3}{\rho^L \nu^2}, \tag{5.2}$$

which is the square of the Galileo number and represents the ratio of the driving buoyancy force to the viscous forces. This is in agreement with Ern et al. (2012), who summarized the

problem of a single bubble rising in a fluid at rest under the effect of buoyancy governed by three parameters: the bubble-to-liquid density ratio (ρ^G/ρ^L), the Archimedes number (*Ar*), and a geometrical parameter relating to the shape of the bubble.
Other effects (different from those occurring with a single bubble rising in quiescent fluid), e.g. swarm effects and the background turbulent shear flow are considered automatically in the DNS velocity fields and no other dimensionless parameters are needed to specify them.
The Archimedes numbers using the non-dimensional parameters of Santarelli & Fröhlich (2015a) (2016b) are reported in Table 5.2. Keeping them the same and using all the other physical dimensional parameters in the r.h.s. of (5.2), yields d_p = 1.456 mm for the small bubbles and d_p = 2.127 mm for the large bubbles (Table 5.2). The ratio d_P/H in Table 5.1 for the different cases then results in the extensions 123.6 mm × 28,0 mm × 61.8 mm of the channel. Here, the channel dimension in the case *Unladen* is the same as the bubble laden cases.

Case	*Unladen*	*SmMany*	*SmFew*	*LaMany*	*BiDisp (Sm/La)*
Ar	-	38171	38171	114528	38171/114528
$d_P(mm)$	-	1.456	1.456	2.1	1.456/2.1
$L_x \times L_y \times L_z\ (mm)$	123.564 × 28 × 61.796				

Table 5.2: The Archimedes number, Ar, the dimensions of the bubble diameter d_P, and the channel in the present work, calculated from the corresponding Ar in the different cases.

5.2 Energy spectra and time scales

The turbulence modulation due to bubbles can be represented by the spectral distribution of the fluctuating energy. Here, the energy spectra evaluated by DNS data are first used to determine an appropriate length and time scale for BIT, which is later used to model the dissipation equation (Section 5.3.2) in a two-equation RANS model. The DNS data as input to obtain the energy spectra was provided by Santarelli & Fröhlich (2015b) based on a collaborated work between the Helmholtz-Zentrum Dresden-Rosssendorf, Germany and Technische Universität Dresden, Germany.
In bubbly flows, Lance & Bataille (1991) were the first to find a power law scaling with a slope of about −8/3 and claimed that it was the feature of the bubble-induced turbulence, in contrast to the classical slope −5/3 for homogeneous isotropic turbulence in single phase flow. They argued the more dissipative spectra due to the presence of the bubbles suggesting a mechanism of eddy disintegration. A general discussion and review regarding to the characteristic slope of BIT until 2005 can be found in Rensen et al. (2005). They proposed a bubblance parameter, *b*, as a dimensionless number to characterize the type of the bubbly flow, defined as

$$b = \frac{\frac{1}{2}\alpha u_r^2}{u_0'^2}, \tag{5.3}$$

where α, u_r, and u_0' are the void fraction, the relative velocity between phases, and the vertical velocity fluctuations in the absence of bubbles, respectively. This is quite close to the suggestion of Lance & Bataille (1991) using the ratio of bubble-induced kinetic energy and kinetic energy of the flow in the absence of bubbles. Here, with large b ($b \gg 1$), the velocity fluctuating of carrier phase is produced rather due to the motion of bubbles, and with small b ($b \ll 1$), the shear-induced turbulence would be more dominant.

In a more recent study, Martinez Mercado et al. (2010) found the scaling of the energy spectrum close to -3 for various gas fractions in the very dilute regime. Their results were obtained by one-dimensional measurements in flows with gas void fractions ranging from 0.8–2.2%, using a phase-sensitive constant temperature anemometry (CTA) probe. In a similar investigation, using PIV in the wake of a bubble swarm, Riboux et al. (2010) found a scaling close to -3 for length scales in order of the bubble diameter o(d_p). Similar results were also measured in homogeneous bubble swarm in a thin gap, which the -3 range located between approximately d_p and $5d_p$ (Bouche, et al., 2014).

The calculation of energy spectra in a bubble-laden flow is delicate, as the phase boundaries tend to disrupt the signal by the passage of bubbles. To overcome the non-continuous velocity signal, different methods in the literature have been used, such as smoothing the discontinuities by a Gauss function (Lance & Bataille, 1991); considering only intervals between bubbles where the velocity signal is continuous (Martinez Mercado et al. (2010), Roghair et al. (2011a), Mendez-Diaz et al. (2013), Bouche et al. (2014), Prakash et al. (2016)); and measuring the wake behind a rising swarm of bubbles, where there is no bubbles (Riboux, et al., 2010).

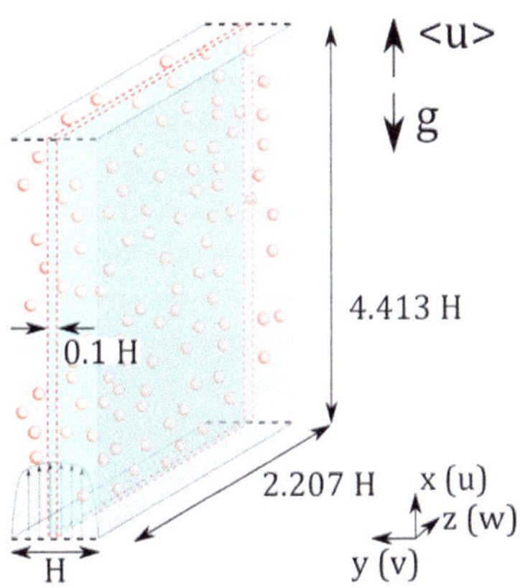

Figure 5.3: Schematic representation of the DNS configuration. The marked region indicates the location where the spatial energy spectra were computed.

Here, a new approach ideally suited for DNS of dilute disperse flows is employed, inspired by the work of Riboux et al. (2010). In their original work, they investigated the flow that develops behind the bubble swarm after the gas injection has been switched off. The spectra of the liquid fluctuations are determined from PIV measurements in the region free of bubbles that immediately follows a swarm of rising bubbles. In the present work, spatial signals of liquid velocity fluctuations in the case *SmMany* were recorded along grid lines in the spanwise direction within the centre region $0.45H < y < 0.55H$ (Figure 5.3). This slab is sufficiently thin to warrant statistical equivalent of these lines. For each line data were recorded whenever the line was free of bubbles resulting in 50,000 one-dimensional data sets, with 2.56×10^7 sample points. As the range in z covers the entire domain, the periodic boundary conditions employed for the simulation result in periodic signals. In Figure 5.4(a), one example is portrayed for an arbitrary instant in time, that the condition mentioned above is fulfilled. The spatial signal of liquid velocity fluctuation in each direction presented in the figure cross over the whole spanwise direction corresponds to 512 sample points due to the resolution of the DNS in the z-direction. Such continuous primal signals do not require windowing, so that they can be Fourier transformed without any problem. This was performed for each velocity component with subsequent averaging over the computed spectra.

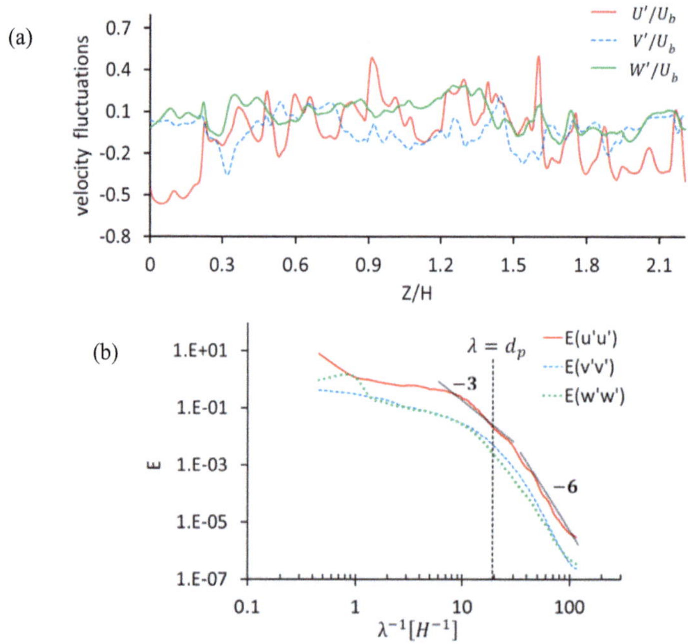

Figure 5.4: (a) Liquid velocity fluctuations in the case *SmMany* along one grid line in the spanwise direction, whenever the line was free of bubbles. (b) Spatial energy spectra of the liquid velocity fluctuations. The dashed vertical line corresponds to the length scale of the bubble diameter in the case *SmMany*.

Figure 5.4(b) shows the resulting energy spectra for the three velocity components. Attention is first focused on the numerical resolution; the finest resolved length scale is 0.0043H, regarding to about 0.08 d_p (see Table 5.1). All the spectra have a similar shape and exhibit three different ranges. A low-wavenumber range resulting from the large-scale background turbulence, a medium range with decay rate around -3 for wave lengths $\lambda = 0.03\text{H}{\sim}0.125\text{H}$, and a dissipation range with a slope around -6. The last part of the energy spectra fall onto the similar slope as the dissipation range for the classic single-phase channel flow (see e.g. by Kim et al. (1987)). This is much steeper than that in Riboux et al. (2010), who found the $-5/3$ scaling recovered in the last range of the spectra and claimed this may be related to a classic Kolmogorov inertial subrange. The difference here may arise from using the windowed FFT in their work, because the selected velocity signals with PIV in their work are not periodic both in vertical and horizontal directions. Using window function is obligated to make sure that the data at the edges are zero to avoid discontinuity. The spectra made by windowed FFT have a common behaviour that an artificial slope closed to -2 is always obtained in the last part of spectra (Bracewell, 1965). Hence, the physics in the last range of their spectra is already removed due to the feature of FFT. Similar results to this issue with an around -2 scaling in the last part of spectra can be found in e.g. Martinez Mercado et al. (2010), Bouche et al. (2014), and Prakash et al. (2016).

The characteristic length scale for BIT is still subject of debate in the literature. Different length scales in the order of d_p in the -3 subrange have been proposed (Martinez Mercado et al. (2010); Riboux et al. (2010); Bouche et al. (2014)) which supports the present result. Comparison between the different studies is complicated by the different methods to calculate the spectra, but the cited works as well as the present one show that d_p is included in the typical -3 subrange for BIT. Hence, from a modelling point of view, d_p is chosen as a starting point for a BIT length scale.

In a similar manner, the choice of a suitable time scale τ is based on temporal spectra. These are obtained in two steps. According to Riboux et al. (2010) and Bouche et al. (2014) the spatial spectra in streamwise direction are the same as those in the spanwise direction. Second, Taylor's hypothesis is applied to convert streamwise spatial spectra into temporal spectra resulting in Figure 5.5(a). Two candidates for a time scale, $(\sqrt{k}/d_p)^{-1}$ (Rzehak & Krepper, 2013) and $(u_r/d_p)^{-1}$ (Troshko & Hassan, 2001) are inserted, together with the single-phase RANS time scale $(\varepsilon/k)^{-1}$ used by Politano et al. (2003) for comparison. Similarly, $\tau = d_p/u_r$ is a good starting point for a time scale characterizing BIT since this frequency is in the typical -3 BIT range, while the others are not. This is consistent with the requirement that a multiscale model for ε has to account for the different length scale and the fast energy decay due to the presence of bubbles (Fox, 2014). Figure 5.5(b) confirms the suitability of d_p/u_r by the experiment of Martinez Mercado et al. (2010) with $u_r/d_p{\sim}40Hz$ and the DNS of Roghair et al. (2011a) with $u_r/d_p{\sim}50Hz$.

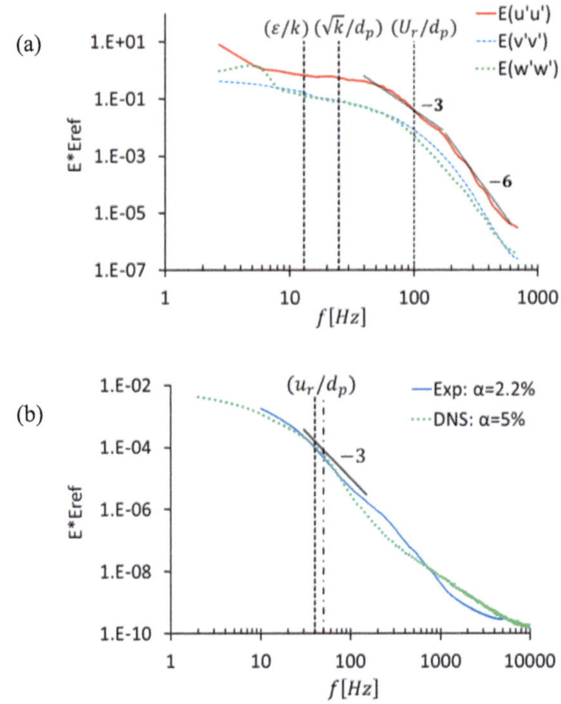

Figure 5.5: (a) Temporal spectra from DNS *SmMany*. (b) Temporal spectra from the experiment of Martinez Mercado et al. (2010) and DNS of Roghair et al. (2011), labelled Exp and DNS, respectively. — — — u_r/d_p from experiment, — · — · — u_r/d_p from DNS.

5.3 RANS modelling of bubbly flow

For incompressible two-phase flow the governing equations of the EE-RANS exhibit a structural similarity as the equations (2.1) and (2.2), where all quantities turn out to be mean values. For convenience, the averaging symbols will herafter be dropped (except where needed to avoid ambiguity), so that all variables using standard letters refer to be mean values. The equations then read

$$\frac{\partial}{\partial t}(\alpha^K \rho^K) + \nabla \cdot (\alpha^K \rho^K \boldsymbol{u}^K) = 0\,, \tag{5.4}$$

$$\frac{D(\alpha^K \rho^K \boldsymbol{u}^K)}{Dt} = -\nabla \cdot (\alpha^K \mu^K \boldsymbol{S}^K) - \alpha^K \nabla p + \alpha^K \rho^K \boldsymbol{g} + \boldsymbol{M}^K - \nabla \cdot (\alpha^K \boldsymbol{\tau}^K_{RANS})\,, \tag{5.5}$$

Established models for the different non-drag interfacial forces are employed here, based on extensive literature study: the model of Auton (1987) for lift, the model of Antal et al. (1991) for the wall force and the model of Burns et al. (2004) for turbulent dispersion. Since the present work focusses on the BIT modelling, the uncertainties resulting from imperfections of the other models employed should be minimized. For this reason the model constants were adapted to reproduce the relative velocity and the gas void distribution of the DNS data. The drag coefficient C_D was determined for each simulation from the balance between drag force, $\boldsymbol{F}_D$, and buoyancy force, $\boldsymbol{F}_B$, according to

$$C_D = \frac{4}{3}\alpha^L \frac{d_p(\rho^L - \rho^G)g}{\rho^L u_r^2}, \tag{5.6}$$

A detailed derivation of this equation is given in Roghair et al. (2011b). The values of C_D in Table 5.1 for each simulation considered are then fixed by an *a priori* approximation using (5.6), where the u_r evaluated from the values employed for the determination of the bubble Reynolds number in different DNS cases, respectively (Table 5.1). Details on how the various coefficients of the non-drag interfacial forces models were determined are given in Section 5.4 below.

5.3.1 Budget of the turbulent kinetic energy

Since the DNS provides the instantaneous liquid velocity fields in three dimensions, it is possible to analyse the exact TKE budget of the liquid phase both in single- and two-phase flows. According to Kataoka & Serizawa (1989), one can write the TKE budget as a two-equation model, one equation for each phase. Since the TKE budgets of DNS provided by Santarelli et al. (2016a) and Santarelli & Fröhlich (2016b) are based on the liquid phase, und furthermore, from the point of view of RANS modelling, that only the turbulence in the liquid phase is treated, in the following, the focus is set on the liquid phase. According to the simplified version of Kataoka & Serizawa (1989) for the modification of TKE by a carried phase, the budgets arise from the various terms contributing to the transport equation of TKE for the liquid phase,

$$\frac{D\overline{\Phi}k}{Dt} = \Pi_k + \varepsilon_k + D_k + S_k \tag{5.7}$$

with the r.h.s. terms written as follows (Kataoka & Serizawa, 1989):

$$production \quad \Pi_k = -\overline{\overline{\Phi u'_i u'_j}}\frac{\partial \overline{\overline{u_i}}}{\partial x_j}, \tag{5.8}$$

$$dissipation \quad \varepsilon_k = -\frac{1}{\rho}\overline{\Phi\tau'_{ij}\frac{\partial u'_i}{\partial x_j}}, \tag{5.9}$$

$$diffusion \quad D_k = -\frac{1}{\rho}\frac{\partial}{\partial x_i}\left(\overline{\Phi p' u'_i}\right) - \frac{\partial}{\partial x_j}\left(\overline{\Phi u'_i u'_i u'_j}\right) + \frac{1}{\rho}\frac{\partial}{\partial x_j}\left(\overline{\Phi u'_i \tau'_{ij}}\right), \tag{5.10}$$

$$interfacial \quad S_k = -\frac{1}{\rho}\overline{p'_L u'_{L,i} n_i I} + \frac{1}{\rho}\overline{\tau'_{L,ij} u'_{L,i} n_j I}, \tag{5.11}$$

with ρ, u and p density, velocity and pressure of liquid phase. In (5.11), p'_L, $u'_{L,i}$, $\tau'_{L,ij}$ are the fluctuations of pressure, i-th velocity component and stress tensor at the liquid side of the phase boundary, respectively. Finally, n_i, I are normal vector at the phase boundary directed toward the gas phase and the interfacial area density. The simple statistical averaging is indicated by a single overbar, while the double overbar denotes the phase-weighted averaging: $\overline{\overline{A}} = \frac{\overline{A\Phi}}{\overline{\Phi}}$, with Φ as an indicator function for the liquid phase defined by $\Phi(x,t) = 1$, if (x,t) within liquid phase and otherwise equal to zero. In the present DNS data, both averaging procedures are performed in the x- and in the z-direction and in time (Santarelli, et al., 2016a). Fluctuations of fluid quantities are defined as $A' = A - \overline{\overline{A}}$, resulting TKE of liquid phase $k = \frac{1}{2}\overline{\overline{u'_i u'_i}}$. The fluctuating viscous stress tensor is $\tau'_{ij} = \nu\left(\frac{\partial u'_i}{\partial x_j} + \frac{\partial u'_j}{\partial x_i}\right)$.

In (5.7), the production term Π_k, dissipation term ε_k and diffusion term D_k are the analogous terms in the single-phase flow as defined in Pope (2000). In this equation, besides the single-phase-like terms, an additional term S_k, the so-called interfacial term represents the interfacial energy transfer between bubbles and liquid and is an additional source term created by the presence of the bubbles (Tanaka & Eaton, 2010). According to Kataoka & Serizawa (1989), the term S_k in (5.7) includes the work which is done by surface tension force and the interfacial drag force. In all the DNS cases considered, the part of work corresponding to surface tension force is zero, since the bubbles are not deformable. So that S_k is equal to the work only due to the interfacial drag force, which is defined as (5.11).

The terms (5.8) – (5.11) were determined explicitly in Santarelli et al. (2016a) and Santarelli & Fröhlich (2016b) for the considered DNS cases. The evaluation of these terms provides gainful insights into the different physical mechanisms that contribute to the TKE. Such terms are the key parameters appearing in closure relations for the RANS modelling, and are described in detail in the following sections.

5.3.2 The modelled *k*-budget and *a priori* test of the BIT term

It is common practice in developing turbulence models for EE-RANS to write the exact TKE transport equation of liquid phase, here, in the framework of two-fluid model. This equation can be derived from (2.2) by evaluating the quantity $k = \frac{1}{2}\overline{u_i' u_i'}$. After manipulation, the result is

$$\begin{aligned}&\frac{\partial}{\partial t}(\alpha^L k) + \overline{u_i}\frac{\partial}{\partial x_i}(\alpha^L k)\\ &\qquad = -\alpha^L \overline{u_i' u_j'}\frac{\partial \overline{u_i}}{\partial x_j} - \frac{1}{\rho^L}\alpha^L \overline{\tau_{ij}' \frac{\partial u'_i}{\partial x_j}} - \frac{1}{\rho^L}\frac{\partial}{\partial x_i}\left(\alpha^L \overline{p' u_i'}\right)\\ &\qquad - \frac{\partial}{\partial x_j}\left(\alpha^L \overline{u_i' u_i' u_j'}\right) + \frac{1}{\rho^L}\frac{\partial}{\partial x_j}\left(\alpha^L \overline{u_i' \tau_{ij}'}\right) + \alpha^L \overline{M_i' u_i'} .\end{aligned} \tag{5.12}$$

One can readily identify the production, dissipation and diffusion terms common to the single-phase TKE equation. The last term accounts for the presence of the dispersed phase and seems similar to (5.11). However, the main point is that the volume-averaged interfacial force included in this term cannot be regarded as fluctuating force in (5.11) at one local position in the phase boundary and it does not represent a valid energy source. Many pervious turbulence closure approaches were made based on (5.12) and were critically reviewed by Crowe et al. (1996) and Crowe (2000). The objective of this section is to propose a physical closure able to describe the BIT effect in the liquid phase.

Here, the SST model of Menter (1994) is employed supplemented with a term accounting for the production of BIT

$$\frac{D(\alpha^L\rho^L k)}{Dt} = \underbrace{\nabla\cdot(\alpha^L(\mu^L+\mu_t^L/\sigma_k)\nabla k)}_{D_k^{RANS}} \underbrace{+\alpha^L P_k}_{\Pi_k^{RANS}} \underbrace{-\alpha^L C_\mu \rho^L \omega k}_{\varepsilon_k^{RANS}} \underbrace{+C_l \boldsymbol{F}_D(\boldsymbol{u}^G-\boldsymbol{u}^L)}_{S_k^{RANS}}, \tag{5.13}$$

where k and ω are denoted without upper index L for simplicity. Further modelling now focuses on S_k, which is a dominant term in the considered cases by Santarelli et al. (2016a) and Santarelli & Fröhlich (2016b). The derivation to the corresponding modelled term S_k^{RANS} is in the departure from the point view of energy balance. Bubbles rise because of gravity and momentum transfer with the surrounding fluid (Mazzitelli & Lohse, 2009). The total energy input by the bubbles is equal to the work done by buoyancy, which is $\boldsymbol{F}_B \cdot \boldsymbol{u}^G$. In the steady state case, this is equal to the work done by the drag, $\boldsymbol{F}_D \cdot \boldsymbol{u}^G$, due to the force balance. This can be decomposed into

$$\boldsymbol{F}_D \cdot \boldsymbol{u}^G = \boldsymbol{F}_D \cdot \boldsymbol{u}^L + \boldsymbol{F}_D \cdot (\boldsymbol{u}^G - \boldsymbol{u}^L). \tag{5.14}$$

If we derive the mean kinetic energy equation by multiplying the momentum equation of liquid phase (5.5) by the liquid mean velocity $\boldsymbol{u}^L$:

$$\frac{\boldsymbol{u}^L \cdot D(\alpha^L \rho^L \boldsymbol{u}^L)}{Dt} = -\boldsymbol{u}^L \cdot \nabla(\alpha^L \mu^L \boldsymbol{S}^L) - \boldsymbol{u}^L \cdot \alpha^L \nabla p + \boldsymbol{u}^L \cdot \alpha^L \rho^L \boldsymbol{g} + \boldsymbol{M}^L \cdot \boldsymbol{u}^L - \boldsymbol{u}^L \cdot \nabla(\alpha^L \boldsymbol{\tau}_t^L). \tag{5.15}$$

Obviously, the contribution $\boldsymbol{F}_D \cdot \boldsymbol{u}^L$ to the term $\boldsymbol{M}^L \cdot \boldsymbol{u}^L$ in the mean kinetic energy equation drives the mean flow, so that the energy converted into turbulence is

$$S_k^{RANS} = C_I \boldsymbol{F}_D \cdot (\boldsymbol{u}^G - \boldsymbol{u}^L)\,, \tag{5.16}$$

with the parameter $C_I \leq 1$, and $\boldsymbol{F}_D = \frac{3}{4d_p} C_D \rho^L \alpha^G |\boldsymbol{u}^G - \boldsymbol{u}^L| (\boldsymbol{u}^G - \boldsymbol{u}^L)$ specified for Euler-Euler approach. Figure 5.6 compares the expression $\boldsymbol{F}_D \cdot (\boldsymbol{u}^G - \boldsymbol{u}^L)$, based on *a priori* tests, to S_k from (5.11). Remarkably, the shape of these curves is very similar. Hence, adjusting the coefficient C_I, bears the potential of successful modelling.

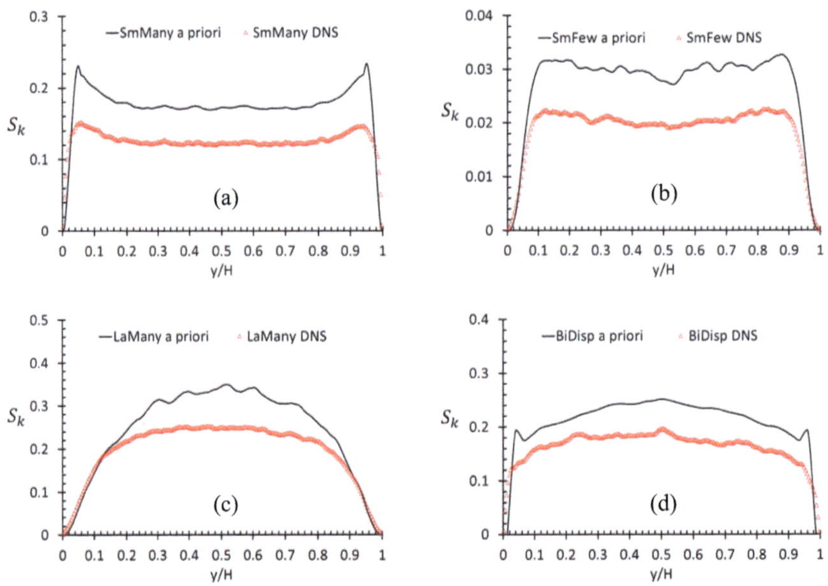

Figure 5.6: Comparison between the interfacial term S_k according to (5.11) and the *a priori* evaluation of S_k^{RANS}/C_I (5.16), both normalized with U_b^3/H. (a) Case *SmMany*; (b) Case *SmFew*; (c) Case *LaMany*; (d) Case *BiDisp*.

5.3.3 Transport equation for ω

Having completed the analysis of the BIT source in the k-equation, the next step focuses on the source term in the modelled ω-equation. The equation of the two-phase SST model is

$$\begin{aligned}\frac{D(\alpha^L\rho^L\omega)}{Dt} &= \underbrace{\alpha^L C_{\omega P}\frac{\rho^L P_k}{\mu_t^L}}_{\Pi_\omega^{RANS}} + \underbrace{\nabla\cdot(\alpha^L(\mu_{mol}^L+\mu_t^L/\sigma_\omega)\nabla\omega)}_{D_\omega^{RANS}} + \underbrace{\alpha^L 2(1-F_1)\rho^L\frac{\nabla k\nabla\omega}{\omega\sigma_{\omega,2}}}_{C_\omega^{RANS}} \\ &\quad \underbrace{-\alpha^L C_{\omega D}\rho^L\omega^2}_{\varepsilon_\omega^{RANS}} + \underbrace{(\frac{1}{C_\mu k}S_\varepsilon^{RANS}-\frac{\omega}{k}S_k^{RANS})}_{S_\omega^{RANS}},\end{aligned} \tag{5.17}$$

where all modelled single-phase terms namely, production term Π_ω^{RANS}, the diffusion term D_ω^{RANS}, the cross-diffusion term C_ω^{RANS}, and the dissipation term $\varepsilon_\omega^{RANS}$ are given by Menter (1994). In (5.17)

$$S_\varepsilon^{RANS} = C_\varepsilon\frac{S_k^{RANS}}{\tau}, \tag{5.18}$$

is the BIT source in the ε-equation transformed to an equivalent source in the ω-equation using the same time scale τ, with C_ε a model parameter. In single-phase flow $\tau = k/\varepsilon$, represents the large-scale eddy turnover time for shear-induced turbulence. But this is far away from the physics of BIT (Figure 5.5a). An essential step now is to replace this time scale with $\tau = d_p/u_r$ characterizing BIT according to Section 5.2 for use in the following.

5.4 Model parameters for the different cases

5.4.1 Computational procedure for EE-RANS

EE-RANS simulations were run in the same domain as the DNS and discretized with $56\times 60\times 51$ grid points in the x-, y-, and z-direction, respectively according to the mesh studies to supply the best prediction. A uniform grid is selected here, considering the different gas distribution in the present DNS cases. Boundary conditions were identical to those of the DNS. At the walls, a no-slip boundary condition is applied for the continuous phase and a free slip condition for the disperse phase.

The liquid and gas are defined as homogeneous distributed at the beginning in the entire computer domain, representing the DNS setup, e.g. with $\alpha^L = 0.9786$ and $\alpha^G = 0.0214$ for the case *SmMany*. Then a constant mass flow rate for *SmMany*, 0.283 kg /s, is calculated from

$$\dot{m} = U_b(\rho^L \alpha^L + \rho^G \alpha^G) A_{yz} \, , \tag{5.19}$$

with $A_{yz} = L_y \times L_z$, being the cross-section of the channel. The same procedure using (5.19) is performed for the other cases as well. For the spatial discretization, a high resolution scheme (Barth & Jesperson, 1989) is employed, and a second-order backward Euler scheme is used in time. The initial condition for the turbulence parameters are considered only for the numerical stability, defined as: liquid turbulent kinetic energy $k = 3.75 \cdot 10^{-7}\, m^2/s^2$ and eddy frequency $\omega = 0.042\, s^{-1}$ (Ansys, 2013).

Before presenting the results of the bubbly flow cases, Figure 5.7 shows the mean liquid velocity and the TKE for the case *Unladen* ($Re_\tau = \frac{u_\tau H/2}{\nu} = 168$, based on friction velocity u_τ and channel half-width $H/2$), performed by standard SST for single-phase flows. A good agreement for mean flow can be observed in Figure 5.7(a) between SST model prediction and DNS data. However, the SST model fails to reproduce the sharp peak of TKE near the walls in the DNS (Figure 5.7b), which is a feature of the *k-ω* model (Wilcox, 1998). This near wall peak is rather passive and has less influence for predicting the mean flow (see Figure 5.7a), since the *ω*-eq. is so calibrated, combining this underpredicted k to provide the correct eddy viscosity or Reynolds shear stress (Menter, 2016). The level of k is slightly overpredicted in the channel centre, similar to the results obtained by Mansour et al. (1988) using a *k-ε* model for a channel flow with the similar Reynolds number ($Re_\tau = 180$).

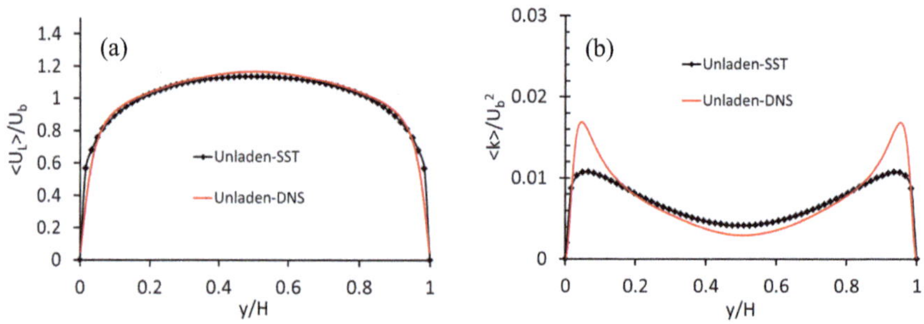

Figure 5.7: Comparison of SST model with DNS data. (a) Streamwise velocity; (b) turbulent kinetic energy.

5.4.2 Procedure for determination of model coefficients

Based on the promising results of Figure 5.6 the idea now is to determine the coefficients C_I and C_ε in (5.16) and (5.18) for each DNS data set, which later on should reveal the general trend. If, however, the relative velocity or the void fraction are different, the BIT model cannot be assessed. Hence, a prerequisite is that the uncertainty created by the other sub-models is removed as much as possible. In order to accomplish this, as mentioned before the drag coefficient was determined from the DNS data directly, i.e. from $u_r = |\boldsymbol{u}^G - \boldsymbol{u}^L|$. Then, an iterative process was carried out by running EE-RANS simulations simultaneously optimizing C_I, C_ε, and C_L, while the other interfacial forces models were employed as they are. The optimization process is schematized in Figure 5.8, from Step 1 to Step 8:

1) C_I and C_ε begin with the values, e.g. $C_I = 1, C_\varepsilon = 1, and\ C_L = 0.5$ for the first run.
2) Possibly bad results for S_k^{RANS}, S_ω^{RANS}, k, and α^G, only u_r is good due to C_D calculated from DNS.
3) In the next simulations, tune C_L to approach α^G of DNS.
4) Tuning C_I and C_ε to let S_k^{RANS} and S_ω^{RANS} fit S_k and S_ω from DNS, respectively.
5) With tuned C_I and C_ε, get the better results for S_k^{RANS} and S_ω^{RANS}. But it is still not good enough, because the gas void fraction changed and not any more perfect (changing C_I and C_ε → changing ν_t → changing modelled turbulent dispersion force $\sim \nu_t$ → changing void fraction profile).
6) Fine tuning C_L again to approach the gas void fraction profiles of DNS.
7) Fine tuning the C_I and C_ε again to get the better results for S_k^{RANS} and S_ω^{RANS}.
8) Repeat the Steps 8 and 9, until void fraction, S_k^{RANS}, and S_ω^{RANS} are all acceptable compared to DNS data.

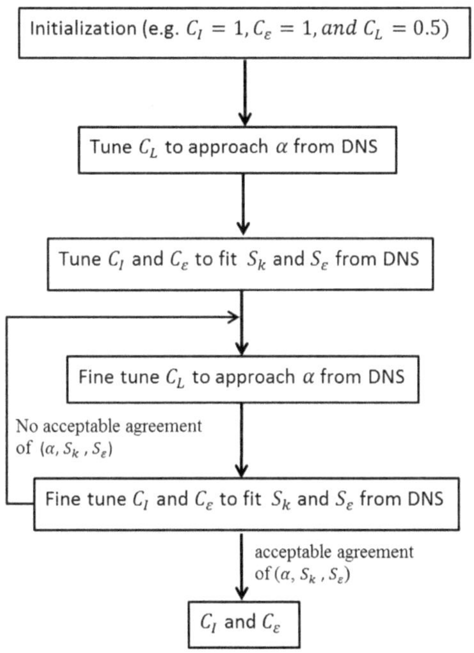

Figure 5.8: Optimization process for determing C_I and C_ε.

5.4.3 Closure for Case *SmMany*

Applying the above procedure it turns out that the resulting value of $C_L = 0.06$, indeed, yields the correct void fraction distribution. The resulting values of $C_l = 0.64$ and $C_\varepsilon = 0.27$ provide very good matches of the DNS data for *SmMany* (Figure 5.9, Figure 5.10, and Figure 5.11). The originality of this process is that the targets of the iterations are not statistics of the mean flow but rather the DNS values of the particular terms to be closed, such as ε_k, S_k, ε_ω, S_ω. The focus here is set on the core region of the channel, (similar as the procedure to fix the model constant in the sink term of the single-phase ε-equation), since close to the walls the situation is more complicated and BIT effect are not dominant. In the centre region, Figure 5.10 shows that the *k*-equation (5.7) reduces to

$$0 \approx \varepsilon_k + S_k \,, \tag{5.20}$$

These terms are both available from the DNS. For ω, no budget is available from the DNS data but, again,

$$0 \approx \varepsilon_\omega + S_\omega \,, \tag{5.21}$$

in the centre for the same reasons as with the *k*-equation. Hence, ε_ω has to be determined indirectly from the the DNS data. For this purpose, it is represented by the expression for $\varepsilon_\omega^{RANS}$ from (5.17), where α^L, ε and k are determined by the DNS data

$$\varepsilon_\omega = -\alpha_{DNS}^L C_{\omega D,2} \rho^L (\omega_{DNS})^2 \,, \tag{5.22}$$

with

$$\omega_{DNS} = \frac{\varepsilon_{DNS}}{C_\mu k_{DNS}} \,, \tag{5.23}$$

using the standard coefficients $C_{\omega D,2} = 0.0828$ and $C_\mu = 0.09$ (Menter, 1994). The equation (5.21) then results in the expression for S_ω, with $S_\omega = |\varepsilon_\omega|$.
Note, as an aside, that the estimations of ε_ω and S_ω here are derived from the point view of solving the modelled turbulence equation to reproduce the turbulence quantities of DNS; they can be, but need not be the exact physical ones.

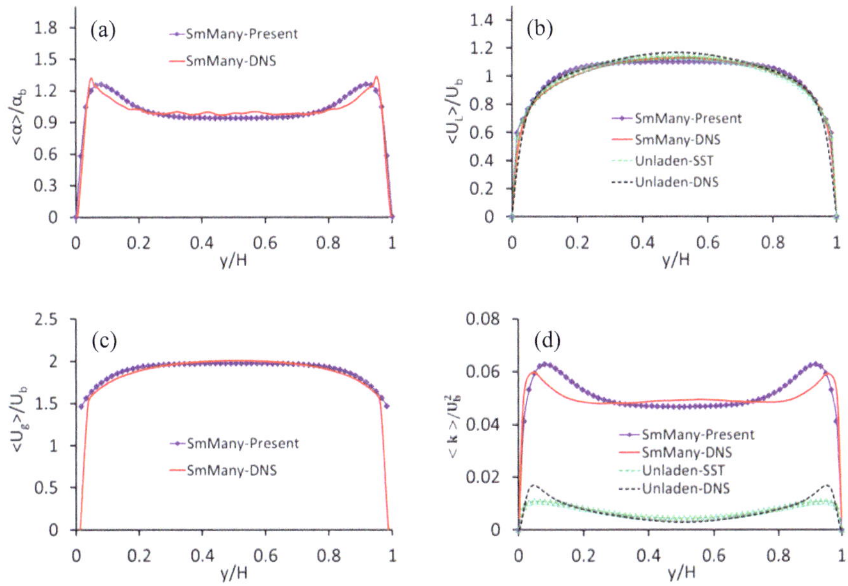

Figure 5.9: One-point statistics from the EE-RANS model for the cases *SmMany* and *Unladen* compared to the DNS data. (a) Gas void fraction; (b) Liquid streamwise velocity; (c) Gas streamwise velocity; (d) Liquid TKE.

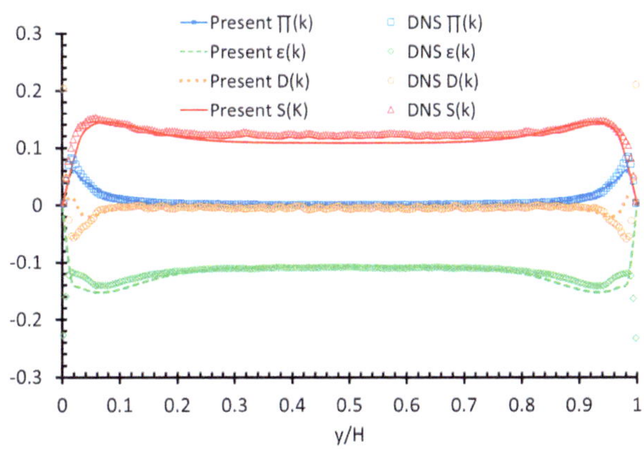

Figure 5.10: Comparison of the present EE-RANS model with the DNS data for the case *SmMany*: *k*-budget with all terms are normalized by U_b^3/H.

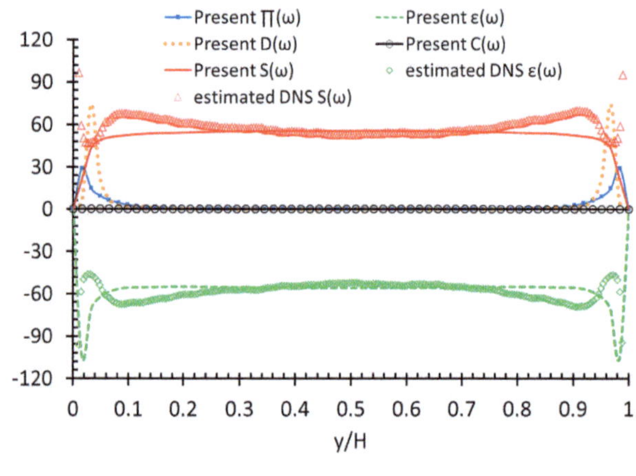

Figure 5.11: Comparison of the present EE-RANS model with the DNS data for the case *SmMany*: ω-budget, all terms normalized by U_b^2/H^2.

After the *SmMany* case is investigated in the previous section, the same procedure to determine the set of parameters $[C_I, C_\varepsilon, C_L]$ is now employed for the cases *SmFew* and *LaMany* (Table 5.1) in the following.

5.4.4 Varying the void fraction

Compared to the case *SmMany,* the case *SmFew* has much smaller number of bubbles with the same bubble diameter (Table 5.1). Figure 5.12 shows the one-point statistics as computed by the present BIT model with the set of the parameters $C_I = 0.66$, $C_\varepsilon = 0.15$ and $C_L = 0.0055$ obtained from the optimization process shown in Figure 5.8.

Figure 5.12(a) shows the mean gas void fraction, which can be reproduced very well in the EE-RANS simulation using the present setting for interfacial force modelling as described in Section 5.3. It demonstrates that in the dilute *SmFew* case, the gas distribution is more uniform across the channel than the case *SmMany*. Here, the different profiles result from the interaction of different forces acting on the bubbles in the considered cases, as extensively discussed in Santarelli & Fröhlich (2015a). The slight differences between the present EE-RANS simulation and DNS can be seen in Figure 5.12(b), depicting the mean liquid velocity. The profile in DNS is flatter than the computed one. This is probably due to the constant C_μ, with the same value 0.09 as in single-phase flow used in the present study, which is too low for this case. The more pronounced peak for mean liquid velocity in the centre leads to a more pronounced peak for the gas velocity profile in EE-RANS compared to the DNS data (Figure 5.12c). The relative velocity shown in Figure 5.12(c) again has the excellent agreement with the DNS data due to the pre-defined C_D (Table 5.1) based on DNS data. Figure 5.12(d) shows the TKE of the liquid phase. The present BIT model gives a very good result due to its ability to match the interfacial terms in the DNS for k- and ω-equation in the channel centre, respectively (Figure 5.13). The underprediction for the production term close to the wall (Figure 5.13a) results from the difference in liquid mean velocity profile (Figure 5.12b).

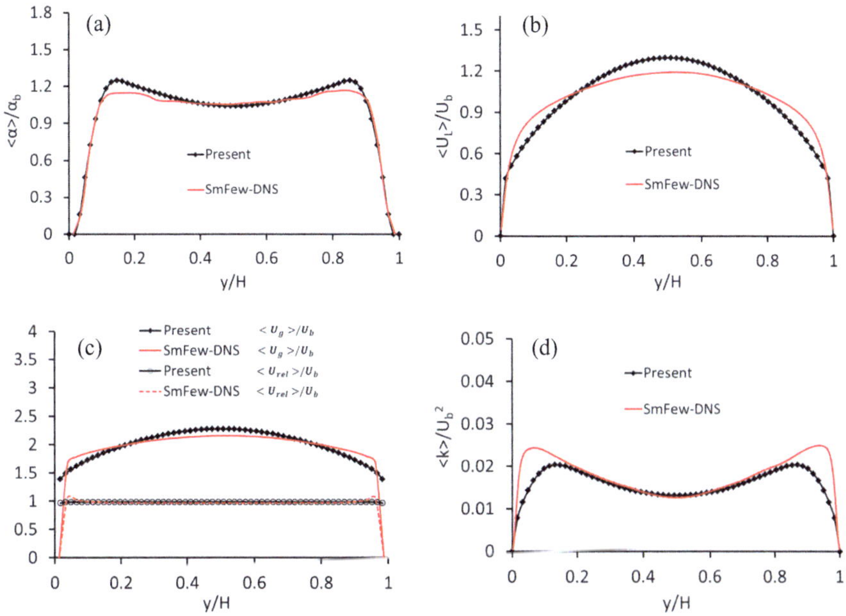

Figure 5.12: One-point statistics from the EE-RANS model for the case *SmFew* compared to the DNS data. (a) Gas void fraction; (b) Liquid streamwise velocity; (c) Gas streamwise velocity & relative velocity; (d) Liquid TKE.

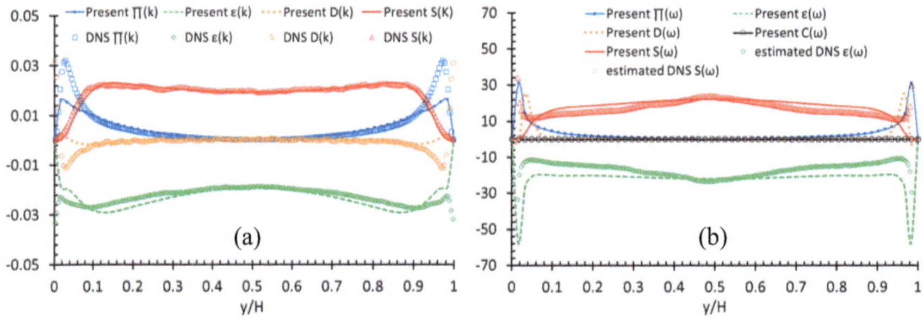

Figure 5.13: Comparison of the present EE-RANS model with the DNS data for the case *SmFew*. (a) k-budget with all terms are normalized by U_b^3/H; (b) ω-budget, all terms normalized by U_b^2/H^2.

5.4.5 Varying the bubble size

The case *LaMany* has the same gas void fraction as *SmMany*, but deals with larger bubbles (Table 5.1). Again, using the optimization process (Figure 5.8) yields [0.75, 0.2, -0.04] for $[C_I, C_\varepsilon, C_L]$. Figure 5.14(a) and (b) show a quite good quantitative agreement between DNS and predicted gas void fraction and liquid velocity, respectively. The mean liquid velocity (Figure 5.14b) is strongly modified by the present case with large bubbles compared to the case *Unladen*. This is different as in the case *SmMany*, where the mean velocity profile only slightly influenced by the bubbles (Figure 5.9b). The shape deviates from the turbulent velocity profile in traditional single-phase channel flows (e.g. by Kim et al. (1987)), with higher values in the centre region due to the higher bubble concentration there (Figure 5.14a) (Santarelli & Fröhlich, 2016b). In the present EE-RANS simulation, the velocity profile obtains a little more pronounced centre peak compared to the DNS data, which is similar to the issue in the case *SmFew* due to the C_μ constant. In a similar manner, this issue then results in a more pronounced gas velocity profile (see Figure 5.14c) obtained using the corresponding C_D (Table 5.1) based on DNS data for the present case. The predicted TKE provides a good agreement with the result obtained by DNS. A slight underprediction in the near wall region can be explained by the difference in the calculated S_ω^{RANS} and $\varepsilon_\omega^{RANS}$with the present model compared to the estimated S_ω and ε_ω from DNS, respectively (Figure 5.15b). This uncertainty can unfortunately not be removed, although the optimization process is employed. The modelled k-source with the parameter $C_I = 0.75$ fits the exact value of the interfacial term in the k-equation (Figure 5.15a).

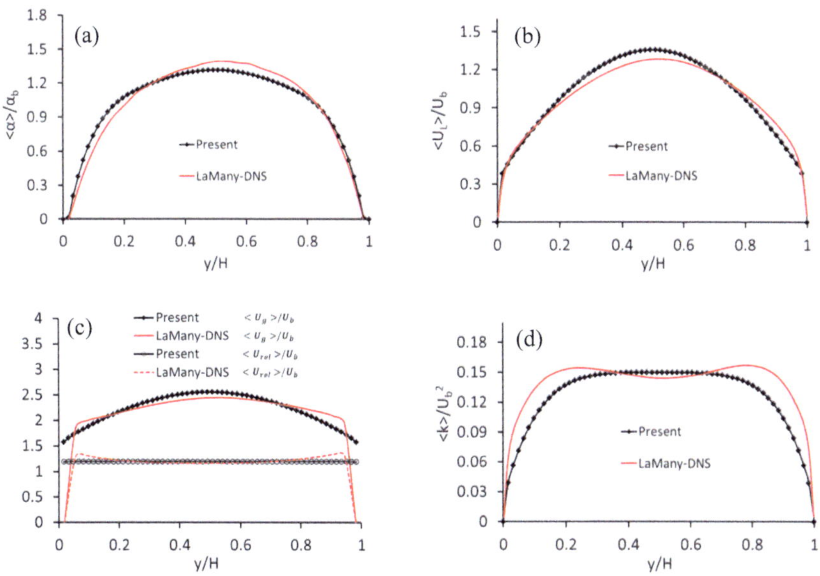

Figure 5.14: One-point statistics from the EE-RANS model for the case *LaMany* compared to the DNS data. (a) Gas void fraction; (b) Liquid streamwise velocity; (c) Gas streamwise velocity & relative velocity; (d) Liquid TKE.

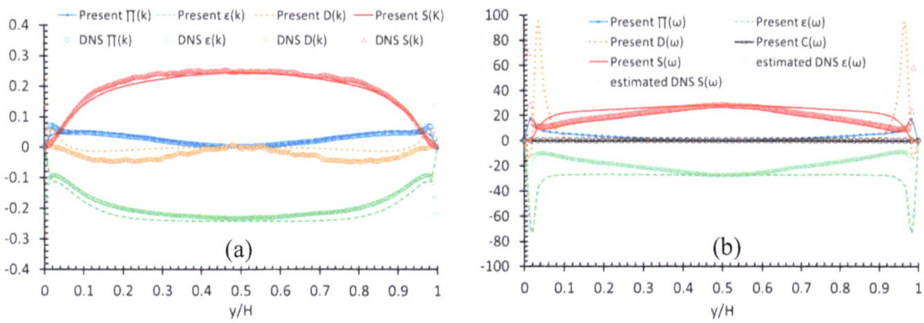

Figure 5.15: Comparison of the present EE-RANS model with the DNS data for the case *LaMany*. (a) k-budget with all terms are normalized by U_b^3/H; (b) ω-budget, all terms normalized by U_b^2/H^2.

5.4.6 Modelling of the bi-disperse bubbly swarm

The previous sections were focussed on mono-disperse bubbly flow. In reality, however, bubble swarms often consist of bubbles with different sizes and have a poly-disperse distribution due to injection conditions, bubble coalescence and break-up, and mass transfer between the phases. For BIT modelling of bi-disperse bubble swarms in the framework of two-equation RANS model, linear superposition of the source term for each gas group is used in the traditional way, both in the TKE and dissipation equations (e.g. by Politano et al. (2003); Krepper et al. (2008), and Ziegenhein et al. (2015)). However, a detailed investigation to validate such an approach in bi-disperse bubbly swarms is not found in the literature. This issue is investigated in the following. The objective here is to extend the BIT modelling for mono-disperse bubbly flow to bi-disperse bubble swarm simulations with the aid of available DNS data.

DNS of bi-disperse bubbly flow were investigated by few authors (Göz et al. (2001), Göz & Sommerfeld (2004); Lu & Tryggvason (2013), and Roghair et al. (2013)). Here, the study of Roghair et al. (2013) deserves special attention. They performed DNS of bi-disperse bubble swarms of deformable bubbles in a periodic domain to extend the drag correlation derived from the mono-disperse bubble swarm to simulate the dense bi-disperse bubbly flow. The influence of the gas fraction, volume fraction ratio and bubble diameter ratio was investigated and compared to the correlation (Roghair, et al., 2011b) for a mono-disperse bubble swarm derived from their previous study using DNS data. As a conclusion, the authors observed that the drag coefficient of the bubbles in bi-disperse bubbly case can be described using the correlation derived from mono-disperse bubble swarms within 15% accuracy.

A detailed investigation of BIT in bi-disperse cases could also be followed by a similar manner to the study of Roghair et al. (2013). However, in the present DNS data, there is only one bi-disperse case, *BiDisp* (see Table 5.1) available. Furthermore, the corresponding interfacial term in the TKE budget is regarded to the summarized contribution of the both bubble sizes, not for the separate one. Nevertheless, this provides the possibility to check the treatment of the BIT source terms in bi-disperse case by the first time, i.e. checking the linear superposition of the source term considering each bubble size. Apart from this DNS data, the author has not found other studies accounting for TKE budgets analysis for the bi-disperse cases for bubbly flows, so far.

The linear superposition of the source terms in the case *BiDisp* are:

$$S_{k,total}^{RANS} = S_{k,Sm}^{RANS} + S_{k,La}^{RANS} \tag{5.24}$$

for k-source and

$$S_{\varepsilon,total}^{RANS} = S_{\varepsilon,Sm}^{RANS} + S_{\varepsilon,La}^{RANS} \tag{5.25}$$

for ε-source, respectively. According to the formulation of the ω-source in (5.17), namely $S_{\omega}^{RANS} = \frac{1}{C_\mu k} S_{\varepsilon}^{RANS} - \frac{\omega}{k} S_{k}^{RANS}$, the source of the ω-equation can also be represented linear, as

$$S_{\omega,total}^{RANS} = S_{\omega,Sm}^{RANS} + S_{\omega,La}^{RANS}, \tag{5.26}$$

Since the Archimedes number does not change, the modification on the bubble Reynolds number and the drag coefficient for each bubble size are small compared to the corresponding mono-disperse cases (*SmMany* and *LaMany*), respectively. It can be expected that the influence of the bubble with the particular bubble size on the liquid fluctuations does not change between the considered cases with the same bubble size. So the parameters C_I and C_ε are directly adapted from the EE-RANS simulations of *SmMany* ($C_I = 0.64,\ C_\varepsilon = 0.27$) and *LaMany* ($C_I = 0.75,\ C_\varepsilon = 0.2$) from the previous sections for the related bubble size in the case *BiDisp*.
Note that the corresponding value of the source terms in the *k*- and ω-equation is very different in the case *SmMany* with small bubbles (Figure 5.13) compared to the case *LaMany* with the large bubbles (Figure 5.15). Hence, it is absolute important to reproduce the profiles of the gas void fraction for small bubbles and large bubbles, respectively, not only the profile of the total gas void fraction in the present simulation. This affords that the simulation should be run in the framework of the inhomogeneous MUSIG model (see Section 3.2.1), that different momentum equations describing the pre-defined gas groups. In particular, here, one group for the small bubbles and the other for the large bubbles.

Case	*SmMany*	*SmFew*	*LaMany*	*BiDisp (Sm/La)*
C_L	0.06	0.0055	-0.03	0.08/-0.04

Table 5.3: The lift coefficient C_L in the present EE-RANS simulations to match the corresponding gas distributions in the different cases.

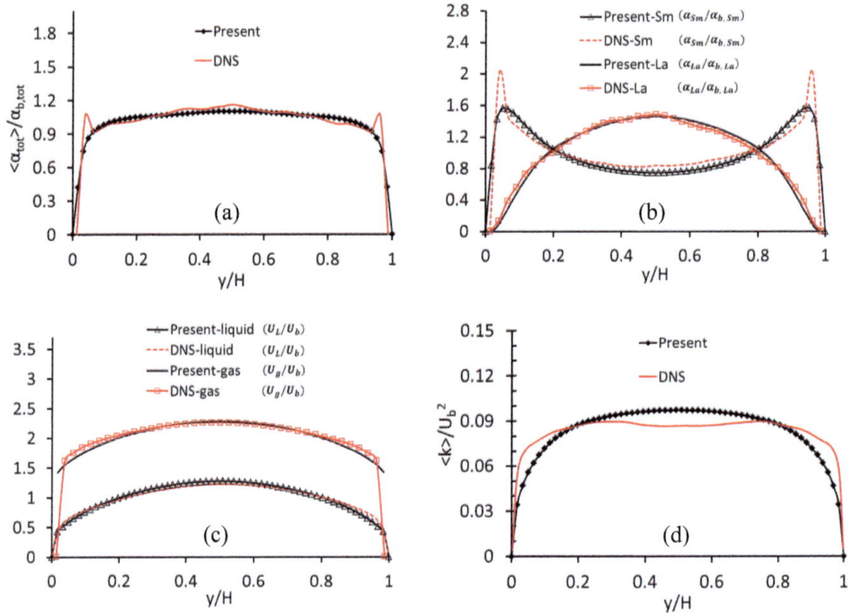

Figure 5.16: One-point statistics from the EE-RANS model for the case *BiDisp* compared to the DNS data. (a) Total gas void fraction; (b) Gas void fraction for small and large bubbles, respectivly; (c) Liquid streamwise velocity & gas streamwise velocity; (d) Liquid TKE.

The interfacial force setting follows Section 5.3. Here, the drag coefficients for small and large bubbles are calculated separately using (5.6) based on DNS data for the each bubble size. The lift coefficients C_L are adjusted to fit the void fraction profiles for small and large bubbles, respectively (Figure 5.16b). For convenience, the corresponding values of C_L are listed in Table 5.3, together with the other three bubble-laden cases. The good agreement in Figure 5.16(b) then results in the well predicted profile of the total gas void fraction (Figure 5.16a). The liquid and gas velocity profiles (Figure 5.16c) can be well reproduced in the present simulation and they lie somehow between the ones of the case *SmMany* and the case *LaMany* as expected.

With the simply adapted C_I and C_ε from the EE-RANS simulations of *SmMany* and *LaMany,* the TKE of the liquid can be determined within 10% accuracy compared to the DNS value (Figure 5.16d). The modelled k-budget and ω-budget are shown in Figure 5.17(a) and (b), respectively, which the k-source in very good agreement with the exact interfacial term from the DNS. Qualitatively good agreement of the ω-source can also be obtained except in the centre region, where it is slight underpredicted. This results in the overprediction for TKE in the channel centre as shown in Figure 5.16(d).

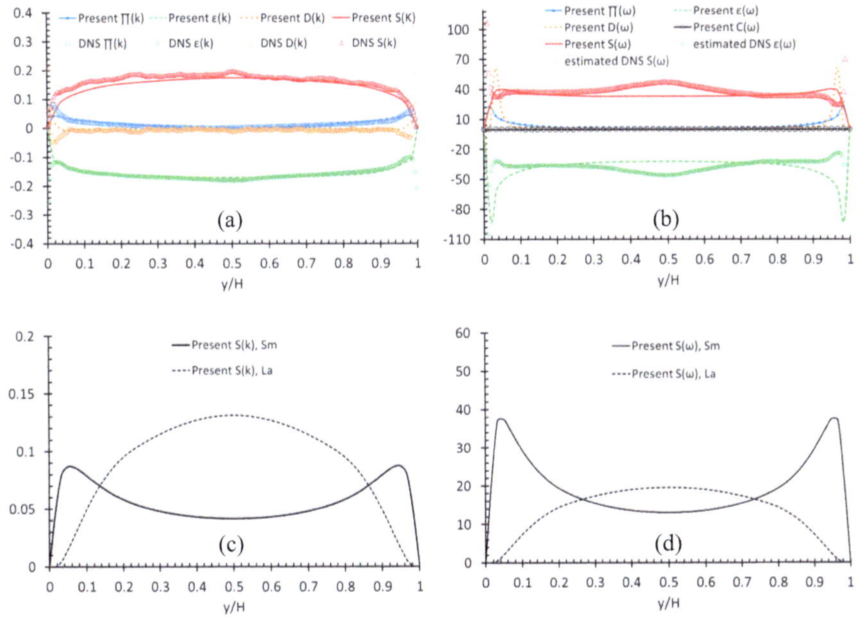

Figure 5.17: Budgets analysis of the present EE-RANS model for the case *BiDisp*. (a) k-budget comparison of the present EE-RANS model with the DNS data for the case *BiDisp*, all terms are normalized by U_b^3/H; (b) ω-budget comparison of the present EE-RANS model with the DNS data for the case *BiDisp*, all terms normalized by U_b^2/H^2; (c) Modelled k-source term for small and large bubbles, respectivly; (d) Modelled ω-source term for small and large bubbles, respectivly.

The decomposed modelled k- and ω-source terms (for small and large bubbles) are shown in Figure 5.17(c) and (d), respectively. All profiles have a shape similar to the corresponding gas distribution of the considered bubble size (see Figure 5.16b), with wall peaks for small bubbles and centre peaks for large bubbles. This is consistent with the findings in the previous sections for the monodisperse cases. It indicates that the related set of C_I and C_ε for each bubble size here are reliable; since they are the prerequisite to achieve a good agreement of the total source terms for the k- and ω-equations (Figure 5.17a and b).
The results of the case *BiDisp* confirm the modelling approach for the source terms in bi-disperse cases, that the contributions from the different bubble sizes can be modelled separately and then added to the total contribution.

5.5 Proposed new BIT model

5.5.1 Correlations for C_I & C_ε

The final step is to propose a general model for the BIT terms. Based on the previous experience, this is done by providing a functional relation for the coefficients C_I and C_ε. The importance of Re_P for BIT and the observation by Erdogan & Wörner (2014) that the TKE increases with Re_p for the same number of bubbles suggest to model C_I as a function of Re_P. The couples of the discrete values of C_I and Re_P (Table 5.4) are displayed in Figure 5.18. For the case *BiDisp*, two data points are plotted for the small bubbles and the large bubbles, respectively. Obviously, C_I increases with Re_p. This supports the hypothesis that bubbles with larger Re_P have stronger wake effects than bubbles with small Re_P, and for even smaller Re_P, the wake eventually disappears (Prosperetti, 2015). Hence, an Re_P-dependent expression for C_I seems suitable for S_k^{RANS}. Curve-fitting of C_I by a power law (Figure 5.18), yields

$$S_k^{RANS} = \underbrace{min(0.18 \cdot Re_p^{0.23}, 1)}_{C_I} \boldsymbol{F}_D(\boldsymbol{u}^G - \boldsymbol{u}^L) \,. \tag{5.27}$$

Here, $C_I \leq 1$ is imposed to fulfil the energy balance discussed with (5.16). Furthermore, C_I should disappear for $Re_P \to 0$.

Case	*SmMany*	*SmFew*	*LaMany*	*BiDisp (Sm)*	*BiDisp (La)*
C_I	0.64	0.66	0.75	0.64	0.75
Re_p	235.5	268.3	475.2	233.6	463.6

Table 5.4: Values of C_I and Re_p for the considered cases.

The value of C_ε could be formulated as a function of S_ε^{RANS}, of the BIT time scale or of the BIT length scale, depending on the problem considered. Inspired by Riboux et al. (2010), who found the characteristic length scale of BIT being related to d_p and the drag coefficient, the values of C_ε (Table 5.5) are plotted over C_D in Figure 5.19. The relation $C_\varepsilon = 0.3C_D$ provides a very good match for data points except in the case *SmFew*. This may be due to the very low void fraction (0.029%), so that the influence of the background flow in this case is not totally negligible. Hence, the following closure is proposed:

$$S_\varepsilon^{RANS} = \underbrace{0.3C_D}_{C_\varepsilon} \frac{S_k^{RANS}}{\tau} , \tag{5.28}$$

with $\tau = \frac{d_p}{U_r}$.

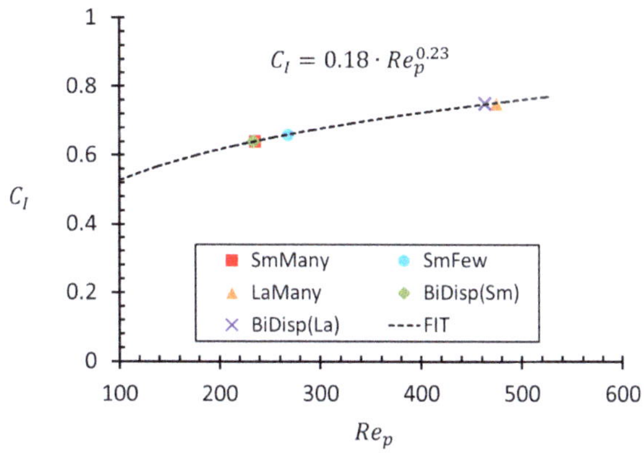

Figure 5.18: Values of C_I as a function of Re_P for all cases considered; Dashed line: Power fit given by $C_I = 0.18 \cdot Re_p^{0.23}$.

Case	*SmMany*	*SmFew*	*LaMany*	*BiDisp (Sm)*	*BiDisp (La)*
C_ε	0.27	0.15	0.2	0.27	0.2
C_D	0.89	0.705	0.666	0.93	0.703

Table 5.5: Values of C_ε and C_D for the considered cases.

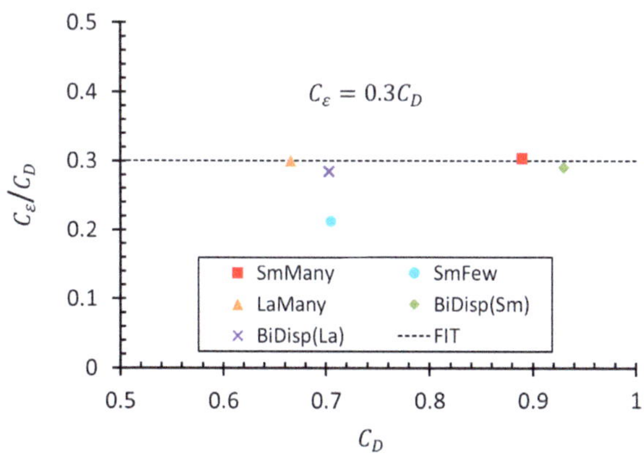

Figure 5.19: Values of C_ε/C_D as a function of C_D for all cases considered. Dashed line: Fit with $C_\varepsilon = 0.3C_D$.

5.5.2 Assessment of existing BIT models

As introduced at the beginning of this chapter, many approaches have been proposed to close the interfacial terms in the framework of EE-RANS. Some detailed report to such closures can be found e.g. by Ilić (2006).

Previous studies related to comparison and assessment of different BIT models can be classified into two categories: the first one is the comparison of the exact interfacial term in the *k*-equation evaluated from the available DNS data with the one achieved by the formula of BIT models, but also using corresponding DNS data (Ilić (2006); Erdogan & Wörner (2014) and Santarelli et al. (2016a)). In the context of LES, such model assessment is known as *a priori* test. The second one is based on *a posteriori* testing, i.e. adapting the BIT models from the literatures in a real EE-RANS simulation, then checking the agreement of the mean flow and the turbulence parameters with the reference data, e.g. by Colombo & Fairweather (2015). However, such studies focus on the comparison of the one-point statistics without more direct analysis of the BIT terms to be closed, since the experimental reference data for such terms are missing due to limitations in measuring techniques (available measuring methods cannot directly evaluate all the quantities in (5.11) for the interfacial term). Therefore, the results can be still confusing and the conclusions were limited in the cases considered. A deeper insight into the predictive features of different BIT models may not be given.

In the present study, *a posteriori* tests are performed in the case *SmMany* using the new proposed model, as well the two representative BIT models from Troshko & Hassan (2001) and Politano et al. (2003). Here, not only the one-point statistics for the case addressed, but also the modelled interfacial terms both in the *k*-equation and in the ω-equation are assessed with the available DNS data. To the best of author's knowledge, such an in-depth comparative study involving both interfacial terms has not been carried out to date. Critical points in the BIT modelling are then discussed in detail.

Table 5.6 tabulates the BIT models used including the corresponding coefficients. In the model of Troshko & Hassan (2001), the k-source is $C_I F_D U_r$, with $C_I = 1$, while the model of Politano et al. (2003) includes a prefactor $4/(3C_D)$. For the ε-source, the both models distinguish from the employed time scale and the corresponding coefficient C_ε. Here, by Troshko & Hassan (2001) the same time scale $\tau = \frac{d_p}{U_r}$ as the present model is used, while Politano et al. (2003) simply adopt the time scale $\tau = k/\varepsilon$, which represents the large-scale eddy turnover time in single phase flow. The value of coefficient, $C_\varepsilon = 1.35 C_D$ by Troshko & Hassan (2001) is a compromise based on reproducing several experimental flow fields considered in their studies. Instead, Politano et al. (2003) used the same coefficient associated with the dissipation term of ε-equation in the single phase k-ε model, reading $C_\varepsilon = 1.92$. However, a detailed explanation for choosing this value is not given by Politano et al. (2003).

	Troshko & Hassan (2001)	Politano et al. (2003)	Present
S_k	$C_I F_D U_r$	$C_I F_D U_r$	$C_I F_D U_r$
C_I	1	$4/(3C_D)$	$0.18 \cdot Re_p^{0.23}$
S_ε	$C_\varepsilon \frac{S_k}{\tau}$	$C_\varepsilon \frac{S_k}{\tau}$	$C_\varepsilon \frac{S_k}{\tau}$
τ	d_p/u_r	k/ε	d_p/u_r
C_ε	$1.35C_D$	1.92	$0.3C_D$

Table 5.6: Summary of the BIT models used for the model comparsion in the present work.

Mean flow statistics

The assessment of the closure models for interfacial terms first recalls the mean flow statistics for the case *SmMany* in Section 5.4.2 to provide a basis for the later discussion. The interfacial forces modelling and the boundary conditions used for the runs addressed by Troshko & Hassan (2001) and Politano et al. (2003) are the same as for the new proposed model reported in Section 5.4.2. All BIT models are embedded in the SST model with the same initial conditions for all parameters. Profiles of gas void fraction, liquid streamwise velocity, gas streamwise velocity, relative velocity, liquid TKE and liquid eddy viscosity are shown in Figure 5.20.

In general, good agreements with the DNS data are obtained for gas void fraction, liquid streamwise velocity, gas streamwise velocity and relative velocity for all simulations. Different gas distributions in Figure 5.20(a) obtained in the near wall region are due to the different modelled forces imposed on the gas phase (discussed later, see Figure 5.21), even if the corresponding coefficients in the interfacial force modelling are identical in all the three simulations. Here, the gas void fraction profile using the BIT model of Troshko & Hassan (2001) has the most pronounced wall peak caused by the lowest magnitude of the modelled turbulent dispersion forces acting on the gas phase close to the walls in the present case; this can be identified in Figure 5.21(c) compared to Figure 5.21(a) (results using the present model) and Figure 5.21(e) (results using the model of Politano et al. (2003)). With the analogy reason, it results in a more pronounced wall peak using the model of Politano et al. (2003) than the one with present model. The difference in gas void fraction close to the walls then causes the somewhat different liquid velocity profiles in the vicinity of $y = 0.1H$ in Figure 5.20(b), where higher velocities are obtained by using the model of Troshko & Hassan (2001) and the one of Politano et al. (2003) compared to the result of the present model, which fits the DNS profile better. This is because the liquid is accelerated by the gas phase, which moves faster due to buoyancy. In particular, more bubbles are accumulated in the near-wall region in the cases using model of Troshko & Hassan (2001) and Politano et al. (2003) compared to the case using the present BIT model. This increases the liquid velocity near the wall, which is responsible for a flatter velocity profile in the former two cases. Note that here the liquid velocity profile in the case *SmMany* is relatively flat in the DNS, i.e. the influence from the eddy viscosity, ν_t, on the mean flow profile is relative small. The shape of the eddy viscosity deviates from the usual consequence in single-phase channel flow, that a higher viscosity corresponding to a flat mean flow profile in the centre and steep slope of mean flow profile near the walls (see the comparison between Figure 5.20b and Figure 5.20f). The gas velocity profiles in Figure 5.20(c) are simply a vertical shifting of Figure 5.20(b) due to the same drag coefficient $C_D = 0.89$ (Table 5.1) involved in all the three simulations, resulting in the same relative velocity between gas and liquid in Figure 5.20(d) for all the cases. The TKE of the liquid depicted in Figure 5.20(e) shows that the result with the present model matches the DNS data better than those with the other two BIT models. For the case run with the Troshko & Hassan (2001) model, the TKE of the liquid is largely underpredicted and a TKE overprediction can be found in the simulation with the model of Politano et al. (2003). The reason is related to the performance of the modelled interfacial terms in the k- and ω-equations, which are discussed further below. Figure 5.20(f) shows the eddy viscosity calculated by the different BIT models and the estimated value using the *a priori* test of the definition of ν_t in k-ε model:

$$\nu_{t,a\,priori} = C_\mu \frac{k_{DNS}^2}{\varepsilon_{DNS}}, \tag{5.29}$$

with $C_\mu = 0.09$. Here, the eddy viscosity unfortunately cannot be related directly to the Reynolds stress, $\overline{u'v'}$, using the gradient hypothesis as

$$\nu_{t,DNS} = \frac{\overline{u'v'}}{\partial u/\partial y}, \tag{5.30}$$

in the traditional way. This is due to the broad plateau of the liquid velocity profile (Figure 5.20b) over the centre region, that in a large distance of the wall normal direction the eddy viscosity according to (5.30) is not defined ($du/dy \approx 0$). There is no wonder, that the present BIT model can fit the estimated eddy viscosity quite well, since the run with this model provides the correct k (Figure 5.20e) and ε (see Figure 5.22, discussed latter) while the others do not.

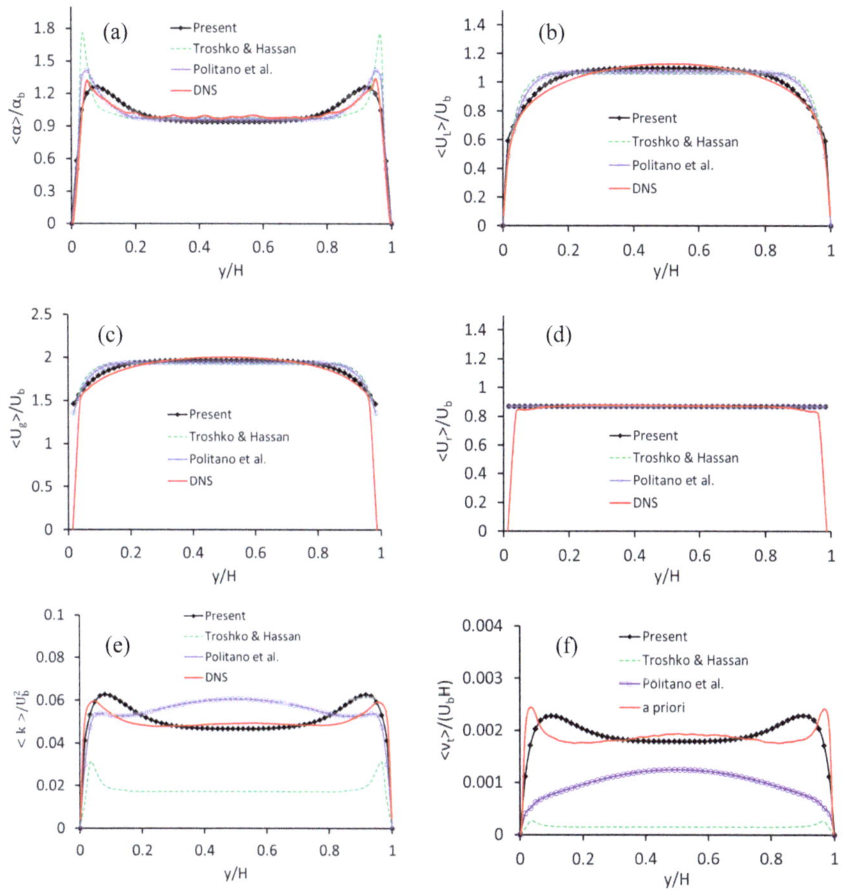

Figure 5.20: Comparison of various BIT models with DNS data for the case SmMany. (a) Gas void fraction; (b) Liquid streamwise velocity; (c) Gas streamwise velocity; (d) Relative velocity; (e) Liquid TKE; (f) Eddy viscosity.

Interfacial force modelling

Accurate interphase force modelling determines the degree of mechanical equilibrium between phases, and then determines the phase distribution across the transversal section of the flow channel (Wang & Yao, 2016). In the present study, due to the strategy of the optimization process as mentioned in Section 5.4 and because no reference data for such forces in the DNS are available, the focus here is not on the assessment of the individual interfacial force modelling, but rather on the influence of a variation of the BIT models on the modelled interfacial forces.

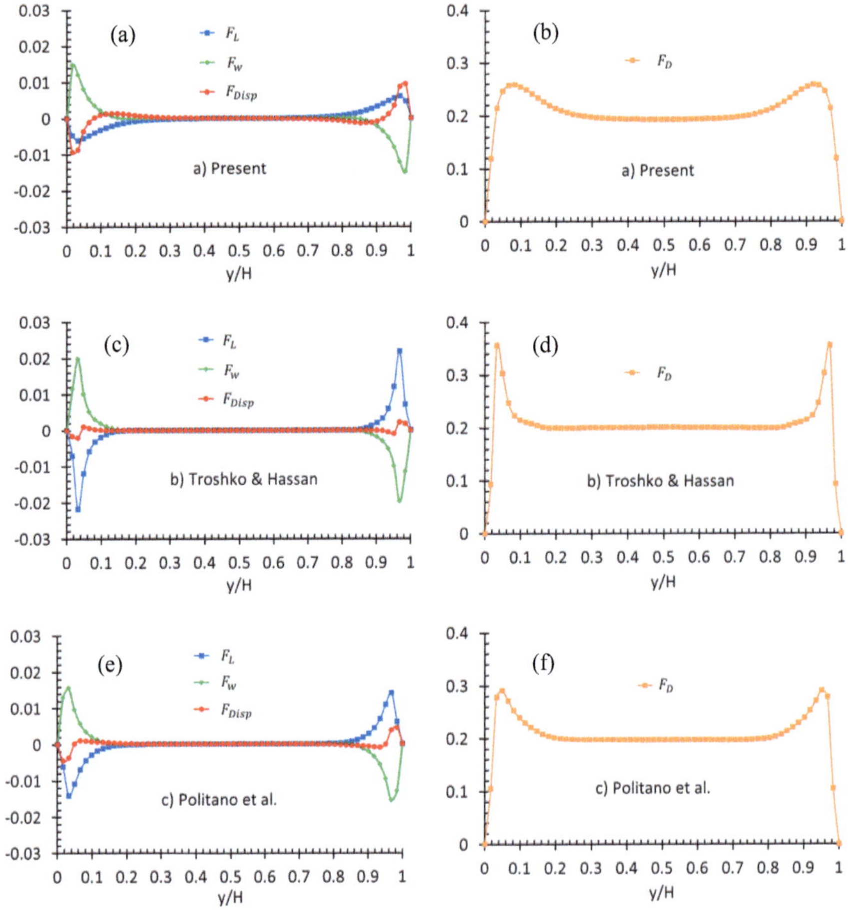

Figure 5.21: Comparison of the modelled interfacial forces using different BIT models for the case *SmMany*. (a) Non-drag forces with the present BIT model; (b) Drag forces with the present BIT model; (c) Non-drag forces with the BIT model of Troshko & Hassan; (d) Drag forces with the BIT model of Troshko & Hassan; (e) Non-drag forces with the BIT model of Politano et al.; (f) Drag forces with the BIT model of Politano et al..

The resulting profiles of the interfacial forces obtained with the different models (based on the same simulations for Figure 5.20) are shown in Figure 5.21. The positive direction is the direction from the wall towards the channel centre. Figure 5.21(a), (c), and (e), depicting the non-drag forces (here, the lift force; the wall force and the turbulent dispersion force), show that they are only active in the near wall region ($0 \leq y/H \leq 0.15$ and $0.85 \leq y/H \leq 1$) and negligible in the channel centre. The sum of the interfacial forces is compensated by the gradient of pressure in the wall normal direction (not shown here). Different lift and wall

forces in Figure 5.21(a), (c), and (e) obtained in the near wall region are due to the differences in the mean flow profiles and the void fraction distribution (see Figure 5.20), even if the coefficients in the corresponding force modelling are identical in all these three simulations. But they are not the source for the different void fraction profiles observed in Figure 5.20(a). Here, the source is the different magnitude of the modelled turbulent dispersion force (2.17). The formula of this modelled force is proportional to the eddy viscosity and the gradient of the void fraction, which prevents a pronounced void fraction peak. Due to the largest eddy viscosity shown in Figure 5.20(f), the present BIT model has the highest magnitude of the turbulence dispersion force close to the wall (Figure 5.21a), trending to drive the bubbles away from the aggregated position in the near wall region. This results in a smoother wall peak of the void fraction using the present BIT model compared to the results obtained by the other two BIT models (see Figure 5.20a). As expected, the Troshko & Hassan model produced the lowest turbulent dispersion force (Figure 5.21c) due to the smallest eddy viscosity, as reflected in Figure 5.20(f), with the Politano et al. model being in the middle.
Figure 5.21(b), (d), and (f) shows the drag force computed by the different BIT models using the same C_D. The shapes in all the three simulations are nearly the same as the corresponding void fraction profiles in Figure 5.20(a), which are the sole difference for evaluating the corresponding drag forces with (2.4). The magnitude of the drag force is about 2 orders of the non-drag forces close to the wall.

k-budgets and ω-budgets

The reason for the different behaviour of the three BIT models can be found in the comparison of the k-budgets and the ω-budgets, which give a deep insight to the models performances.
Figure 5.22 compares the k-source term and the dissipation term in the k-equation, which are the dominating terms in the case *SmMany*. For the k-source, except the present model, both BIT models of Troshko & Hassan (2001) and Politano et al. (2003) overpredict the values evaluated in DNS, about 45% and 110%, respectively. Figure 5.23 shows a comparison between the modelled ω-budgets and the estimated DNS ω-budgets, again, only for the dominating terms (ω-source and the dissipation term in the ω-equation). The models of Troshko & Hassan (2001) and Politano et al. (2003) fail to capture the correct level of the ω-source; especially the value evaluated by the Troshko & Hassan is surprisingly around 2 orders of the estimated value of DNS, even if the time scale is identical compared to the present model and the only difference in the expression of ω-source is: $C_\varepsilon = 1.35C_D$ instead of the present model with $C_\varepsilon = 0.3C_D$. This indicates the large sensitivity of C_ε for the modelled ω-source, even if the same time scale in the source term is used. The Politano et al. model has a single phase time scale k/ε for the ω-source, which the corresponding value is much larger than the time scale d_p/u_r (Figure 5.5a). This model overpredictes the ω-source about 260% (see Figure 5.23).
The incorrect prediction of the source terms in the k- and ω-equation using the BIT models of Troshko & Hassan (2001) and Politano et al. (2003) gives the direct reason that both models fail to capture the TKE in the considered case, as was reflected in Figure 5.20(e). Noting that although the overprediction of the k-source for the Troshko & Hassan model, the calculated TKE is much smaller than that in the DNS. This is due to the dramatic overprediction for the ω-source in this model, which induces a huge value of ω. But the product $\omega \cdot k$ represents the

dissipation term in the *k*-equation and compensates the *k*-source, which has a much smaller rate of overprediction compared to the ω-source.

Such a finding indicates that a smaller overprediction of the ω-source would give a good agreement for TKE, even if both source terms in *k*- and ω-equations are overpredicted. This issue is discussed in the next section in detail.

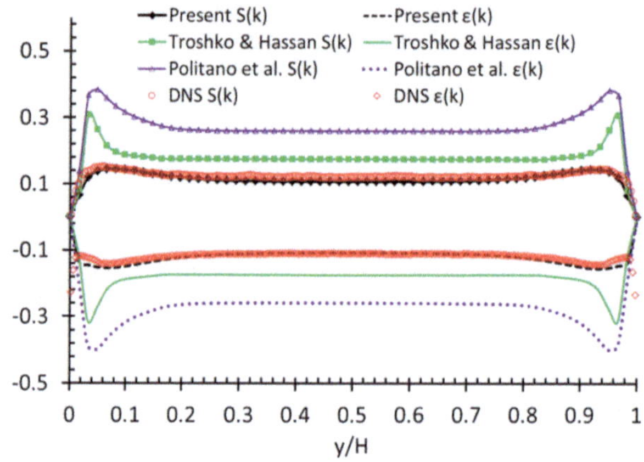

Figure 5.22: Comparison of the interfacial term and dissipation term in the budget of turbulent kinetic energy using various BIT models with DNS data in the case *SmMany*.

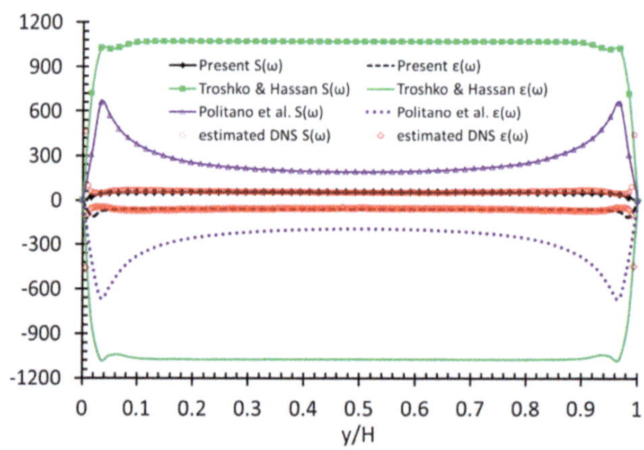

Figure 5.23: Comparison of the interfacial term and dissipation term in the budget of turbulent eddy frequency using various BIT models with DNS data in the case *SmMany*.

5.6 Critical points in BIT modelling

In the following, two critical points with regarding the previous BIT proposals are discussed basing on the case *SmMany* as an example.

5.6.1 BIT modelling based only on analysis of TKE budget

Detailed *a priori* tests of the different approaches for k-source are presented by Ilić (2006), Erdogan & Wörner (2014) and Santarelli et al. (2016a) in the framework of comparing the modelled TKE budget with the ones obtained by DNS in bubbly flows. Such studies provide the direct information about the relative importance of each term in the TKE equation in a particular flow considered. Then, a feedback after comparison is given with regarding to these terms, which need to be improved. However, the *a priori* testing suffers from the general limitation similar as that in LES (Meneveau, 1994), and it cannot be guaranteed that their recommendations necessarily result in improved performance in an actual simulation (Bou-Zeid, 2015). For instance, mean void fraction and velocity profiles are influenced by the BIT model itself as it is part of the governing equations and, therefore, affect the solution for the mean quantities. Since this feedback is non-linear, *a posteriori* tests of any potential model in Euler-Euler simulations are necessary (Wörner & Erdogan, 2013).

Furthermore, even the exact TKE budget can be reproduced with the aid of a proposed k-source in a real simulation; it does not necessarily mean that the resulting TKE should be correct, if the BIT term in the dissipation equation is not considered. Here, the agreement with the TKE budget of DNS can only assure that the modelled turbulent dissipation and interfacial terms in the k-equation are the same as those in the DNS, but not the TKE. This issue can be demonstrated with a simple simulation, labelled RANS1 performed for the case *SmMany,* using $C_I = 0.64$ for S_k^{RANS} and $C_\varepsilon = 0.5$ for S_ε^{RANS} (Figure 5.24), all the other settings are identical to those in Section 5.4.2. Here, $C_I = 0.64$ is adopted from Table 5.4 to match the exact dissipation and interfacial term (Figure 5.24a). $C_\varepsilon = 0.5$ for S_ε^{RANS} is chosen arbitrarily to differ the value $C_\varepsilon = 0.27$ suggested in Table 5.5 for the case *SmMany*. Obviously, an overprediction of the modelled ω-source can be observed (Figure 5.24d). The void fraction is not influenced by this effect; a good agreement can be still achieved as shown in Figure 5.24(b), which hints that for the considered case, turbulence does not influence the gas distribution too much. This is important to identify the reason for the underpredicted TKE in Figure 5.24(c), that is nearly only from the overprediction of the modelled ω-source and the influence of other factors are negligible.

In fact, in two-equation RANS models, the task of the modelled dissipation equation (ε- or ω-equation) is combined with the modelled k-equation to provide the correct k and the turbulent dissipation. The example confirms that the same procedure as the analysis of the TKE budget must be performed in a similar manner for the ω-budget as introduced in Section 5.4.2.

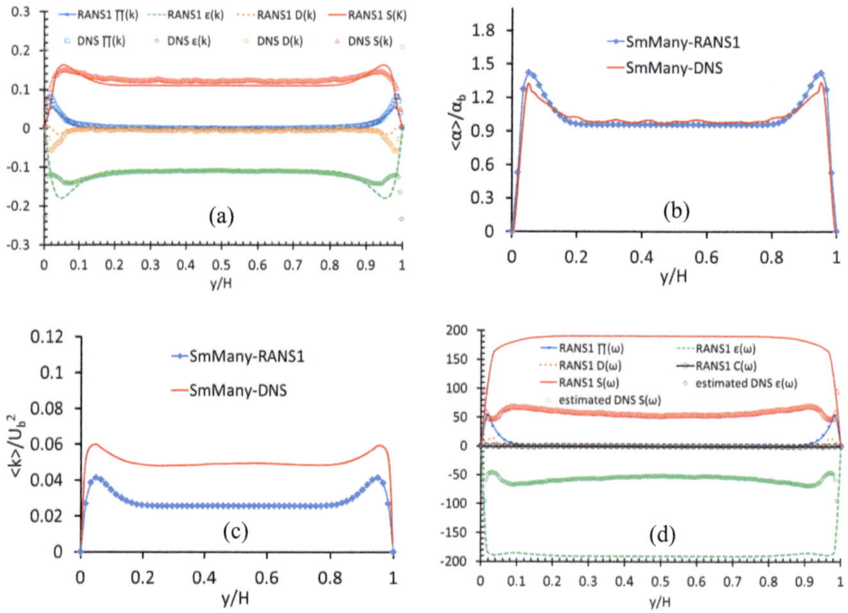

Figure 5.24: Comparison of a test case (RANS1: $C_I = 0.64$ and $C_\varepsilon = 0.5$) with DNS data for the case *SmMany*. (a) k-budget; (b) Gas void fraction; (c) Liquid TKE; (d) ω-budget.

5.6.2 BIT modelling based on fitting TKE

It has been the author's experience that tuning the constants in the source terms can lead to a significant change of the BIT model performance for simulating the BIT dominated bubbly flows. Many previous BIT studies used *ad hoc* adjustments of the model constants in the source terms to reproduce the flow fields in the considered reference data. Examples for such kinds of BIT models can be found in the review of Ilić (2006) until 2006, and the more current studies e.g. given by Colombo & Fairweather (2015) and Islam et al. (2016).
However, the study in Section 5.4 implies that such model constants may provide pitfalls. This point can be supported with a simple simulation, labelled RANS2 for the case *SmMany*, using $C_I = 0.4$ and $C_\varepsilon = 0.17$ instead of the ones ($C_I = 0.64$ and $C_\varepsilon = 0.27$) from Table 5.4 and Table 5.5 for k- and ω-source, a good fit of TKE profile can also be achieved (Figure 5.25a). Here, for the run of RANS2, all the other settings are identical as in Section 5.4.2 and the calculated profile of gas void fraction has a good agreement with the DNS data (Figure 5.25b).
Here, the fact that the TKE is well reproduced does not necessarily mean that S_k^{RANS} and S_ε^{RANS} are modelled correctly. Checking closely, in Case RANS2, the good agreement in TKE of *SmMany* with the set $[C_I = 0.4$ and $C_\varepsilon = 0.17]$ results from underpredicting the source

term in both the k- and the ω-equation (see Figure 5.25c and d). This can yield further errors and does not solve the modelling problem.

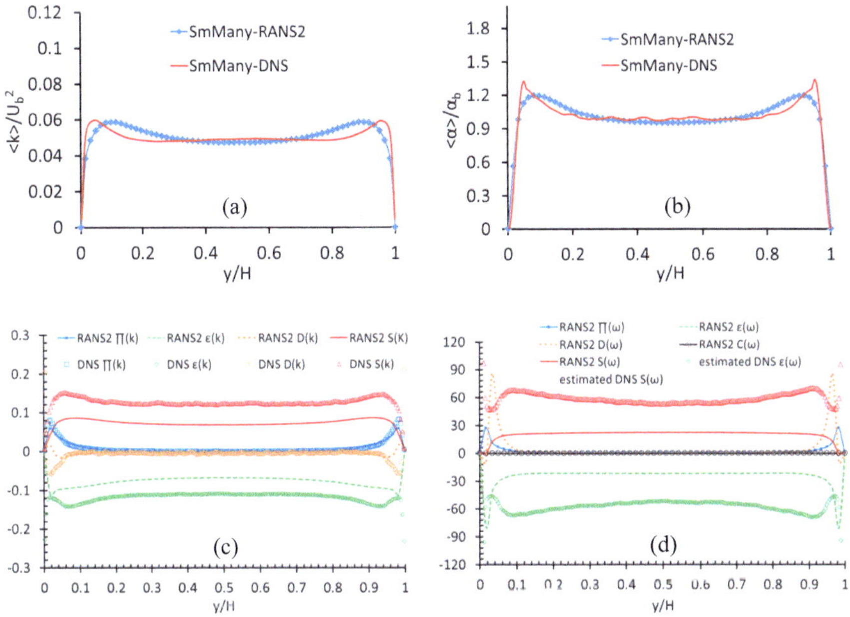

Figure 5.25: Comparison of a test case (RANS2: $C_I = 0.4$ and $C_\varepsilon = 0.17$) with DNS data for the case SmMany. (a) Liquid turbulent kinetic energy; (b) Gas void fraction; (c) k-budget; (d) ω-budget.

From the mathematical point of view, in such two-equation RANS model with dominating BIT terms, there are infinite sets of C_I and C_ε to reproduce the correct TKE, without checking the turbulent dissipation. This can be achieved, if the corresponding S_k^{RANS} and S_ε^{RANS} are both overpredicting or both underpredicting the excat interfacial terms with a suitable rate. It implies one important point with regarding to the strategy of BIT modelling, that such model constants are not correct and cannot be served for the BIT model development.

5.7 First validation of the new BIT model

The previous sections examined various aspects of the new BIT model and made comparisons to other models. In this section, the computed results using the new BIT model are compared to experimental data. The configuration is a rectangular water/air bubble column with a homogeneous gas injection at the bottom (Akbar, et al., 2012).

The experimental setup of Akbar et al. (2012) has been introduced in Section 3.2. Here, the case with smaller gas superficial velocities 3 mm/s (Case 1 in Section 3.2) is considered, since the BIT effect is dominant there, so that the case is very suitable for such a validation. Concerning comparison for turbulence parameters in the framework for EE-RANS, the experimental values of k are estimated, since in the experiment of Akbar et al. (2012), only the velocity fluctuation in the vertical direction was measured for the corresponding measurement line (see Figure 3.1). For the centre of the measurement line, the ratio of the Reynolds normal stresses $\overline{w'w'} : \overline{u'u'} \sim 2.5 : 1$ was observed in the experiment (Akbar, et al., 2012). A suitable approximation yields the ratio of the Reynolds normal stress in the streamwise direction (z) to the horizontal direction (x) equal to 2.5 for the complete measurement line due to the homogeneous gas distribution over the entire line (except the near wall region) (Figure 5.26a). The x-directional and y-directional Reynolds normal stress can be considered equal, so that

$$k \approx \frac{1}{2}\left(\overline{w'w'} + 0.4\overline{w'w'} + 0.4\overline{w'w'}\right) = 0.9\overline{w'w'}\,. \tag{5.31}$$

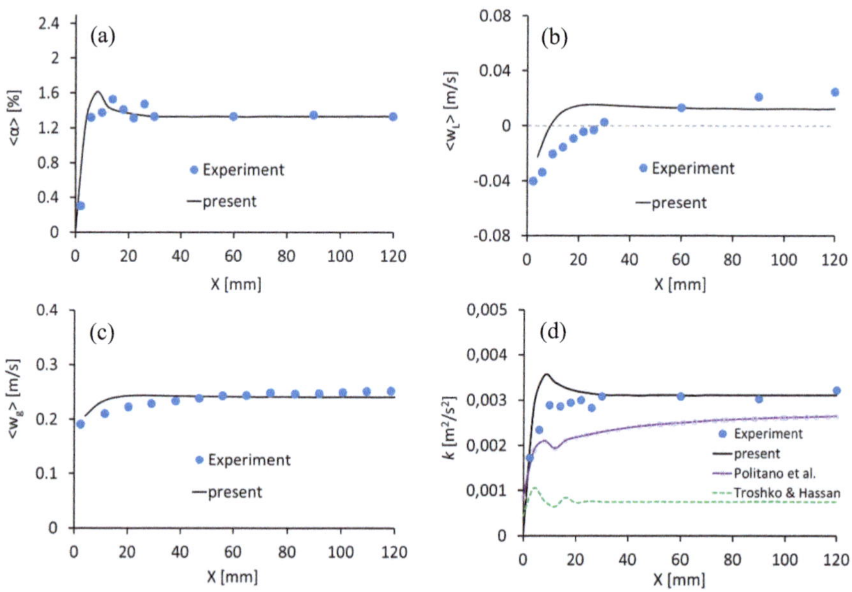

Figure 5.26: Comparison of the present model with experimental data. (a) Gas void fraction; (b) Liquid streamwise velocity; (c) Gas streamwise velocity; (d) Liquid TKE.

The present EE-RANS simulation is performed on an identical grid as that in Section 3.2 for EE-LES. A similar set of interfacial force modelling compared to the settings in the EE-LES are employed as well. The new BIT model is implemented in the SST model of two-phase version.

Figure 5.26 (a), (b), and (c) show a comparison of the gas distribution, vertical liquid velocity, and vertical gas velocity, respectively. The results are comparable to these performed with EE-LES before (see Figure 3.5) and achieve a good agreement with the experimental data. The more flatter profile for liquid velocity in Figure 5.26(b) may due to the inaccuracy in the gas void fraction close to the wall (Figure 5.26a). Figure 5.26(d) shows a comparison of the simulated and experimental liquid TKE. The new BIT model provides superior results and the performance of the model is substantially better than with the models from Troshko & Hassan (2001) and Politano et al. (2003) under exactly the same conditions.

6 Conclusions and discussion

Euler-Euler (EE) simulations have been carried out to investigate the various turbulence modelling approaches for disperse bubbly flows. The considered approaches contain two kinds of scale resolving simulations (LES & SAS) and the traditional RANS simulations.

EE-LES have been performed for the rectangular bubble columns with two distinct designs. The first one corresponds to a homogeneously distributed gas inlet at the bottom of the bubble column (Akbar, et al., 2012). The results obtained with the EE-LES approach reproduce the measured gas volume fraction and liquid velocity profiles in the same way as EL-LES (Akbar, et al., 2012) and EE-URANS (Ziegenhein, et al., 2015). Large improvement can be achieved with the EE-LES method for the turbulence prediction in the case with a higher gas superficial velocity. A near-wall peak in the velocity fluctuation can be reproduced. Here, the approach is successful, as the largest and most energetic scales of motion (comparable in size to the whole domain) are much stronger than the BIT, which are not resolved with this approach. For lower gas superficial velocity, LES may not be the best option for turbulence prediction, since large-scale turbulence is not present and the fluid motion on the smaller scales with the low superficial gas velocity is dominated by the bubble-induced turbulence. Consequently, an improvement for this case cannot be achieved with the present EE-LES approach due to the lack of physical modelling for such small scale fluctuations. Furthermore, a method for estimating the SGS kinetic energy is introduced and investigated. Considering this unresolved part may improve the prediction of the velocity fluctuation. A similar power law like experimental energy spectrum might not be reproducible with EE-LES, since the bubbles are not resolved in Euler-Euler approach and the frequency information related to the bubble wake is lost.
The other demonstration case of EE-LES was for a bubble plume in a flat rectangular bubble column, where large-scale eddies exist. The simulations results were compared with the corresponding experimental data from Pfleger et al. (1999) and good qualitative agreement with the experiment can be achieved for the instantaneous bubble plume shape in general. For the time-averaged liquid velocity profiles at different heights of the column the agreement was quite good, quantitatively. The Reynolds stresses determined with LES are similar in tendency as those obtained from the similar rectangular bubble column experiments of Mudde et al. (1997) and Deen et al. (2000).

The second kind of scale resolving simulation considered in the present study was SAS, where a physical transport equation for turbulent length-scale is employed. The simulation results obtained with SAS for a square cross-sectional bubble plume show good agreement with the LES results and the experiment, both for the instantaneous bubble plume shape and for the mean-flow properties. The turbulence quantities determined with SAS are similar in quality as those obtained from LES in this case. Especially the trend of the fluctuations in the mean flow direction is well captured. SAS hence might be a promising option to be used with the Euler-Euler approach to predict bubbly flows.
The BIT model of Sato was investigated in SAS and LES. It is found, that the increased effective viscosity resulting from this model damps the bubble plume dynamics. This causes a

more pronounced peak in the centre of the liquid velocity profile compared to experimental one. Such an effect due to the Sato BIT model was not found in the Akbar case with a homogeneously distributed gas inlet. One reason could be the flatter gas volume fraction profile in such a bubble column. The impact of the increased effective viscosity caused by the Sato model for this reason is much higher in the bubble plume cases.

The analysis of the TKE budget from direct numerical simulation data can be used for the development of new two-phase RANS models. In the present work, Euler-Euler modelling of vertical bubbly channel flow is performed using the Shear Stress Transport model. In this flow, the BIT is dominating, i.e. the interfacial term compensates nearly the total dissipation in the TKE equation. New coefficients for the source terms in the k-equation and the ω-equation are derived from DNS data. With these improved source terms in the RANS model, the results show a good agreement with the reference data. Furthermore, a new BIT proposal, equations (5.27) and (5.28), for closing the BIT terms in a typical RANS model is constituted. These correlations are expressed in terms of Re_P and C_D, which are available in any EE-RANS simulation. The model is ready for use and can as well be employed for any similar two-equation RANS model.
With a fairly coarse grid, the EE-RANS simulation using the new BIT model can be obtained with relatively small computer power. For instance, in the case *SmMany*, a single run took only 10 hours on 12 cores of a PC cluster. For comparison, the corresponding DNS took 300,000 CPU-hours, using typically 256-512 cores (Santarelli & Fröhlich, 2015a).

"Turbulence and multiphase flows are two of the most challenging topics in fluid mechanics, and when combined they pose a formidable challenge" (Balachandar & Eaton, 2010). The turbulence problem of bubbly flows is sure far from solved, and open issues were addressed throughout the present study. For EE-LES, a future work could consider the approach from Riboux et al. (2013) for single-phase flow, by imposing a localized momentum forcing term in Euler-Euler equations to represent the BIT effect. Regrading to the new developed BIT model for EE-RANS, the proposed model can as well be employed for any similar two-equation RANS model. As such, the model is valid for small bubbles in contaminated fluid, which is highly relevant for practical applications. Further tests should focus on the performance of the model over a wider range of bubble Reynolds numbers and in more complex flow fields. Another key aspect for future research is the prediction of the full Reynolds stress with account for its anisotropy in bubbly flows. The development of such a second-moment closure (Ma, et al., 2017b) for BIT can follow the similar procedure for two-equation RANS model as done in Section 5.

Bibliography

Adoua, R., Legendre, D. & Magnaudet, J., 2009. Reversal of the lift force on an oblate bubble in a weakly viscous linear shear flow. *Journal of Fluid Mechanics,* Volume 628, pp. 23-41.

Akbar, M., Hayashi, K., Hosokawa, S. & Tomiyama, A., 2012a. *Bubble Tracking Simulation of Bubble-induced Pseudo Turbulence.* Kumamoto, 6th Japanese-European Two-Phase Flow Group Meeting.

Akbar, M. H., Hayashi, K., Hosokawa, S. & Tomiyama, A., 2012. Bubble tracking simulation of bubble-induced pseudoturbulence. *Multiphase Science and Technology,* 24(3), pp. 197-222.

Ansys, 2013. *CFX 14.5 Manual: CFX-Solver Modeling Guide..* Canonsburg: Ansys, Inc..

Antal, S., Lahey, R. & Flaherty, J., 1991. Analysis of phase distribution in fully developed laminar bubbly two-phase flow. *International Journal of Multiphase Flow,* Volume 17, pp. 635-652.

Auton, T. R., 1987. The lift force on a spherical body in a rotational flow. *Journal of Fluid Mechanics,* Volume 183, pp. 199-218.

Balachandar, S. & Eaton, J. K., 2010. Turbulent Dispersed Multiphase Flow. *Annual Review of Fluid Mechanics,* Volume 42, pp. 111-33.

Barbosa, C., Legendre, D. & Zenit, R., 2016. Conditions for the sliding-bouncing transition for the interaction of a bubble with an inclined wall. *Physical Review Fluids,* Volume 1, 032201(R).

Barth, T. & Jesperson, D., 1989. The Design and Application of Upwind Schemes on Unstructured Meshes. *AIAA Paper,* Volume 89, 0366.

Biesheuvel, A. & van Wijngaarden, L., 1984. Two-phase flow equations for a dilute dispersion of gas bubbles in liquid. *Journal of Fluid Mechanics,* Volume 148, pp. 301-318.

Bolotnov, I. A. et al., 2011. Detached direct numerical simulations of turbulent two-phase bubbly channel flow. *International Journal of Multiphase Flow ,* 37(6), pp. 647-659.

Bose, S., Moin, P. & You, D., 2010. Grid-independent large-eddy simulation using explicit filtering. *Physics of Fluid,* Volume 22, 105103.

Bouche, E., Roig, V., Risso, F. & Billet, A., 2014. Homogeneous swarm of high-Reynolds-number bubbles rising within a thin gap. Part 2. Liquid dynamics. *Journal of Fluid Mechanics,* Volume 758, pp. 508-521.

Bou-Zeid, E., 2015. Challenging the large eddy simulation technique with advanced a posteriori tests. *Journal of Fluid Mechanics,* Volume 764, pp. 1-4.

Bracewell, R., 1965. The Fourier Transform and its applications. *McGraw Hill.*

Bunner, B. & Tryggvason, G., 2003. Effect of bubble deformation on the stability and properties of bubbly flows. *Journal of Fluid Mechanics,* Volume 495, pp. 77-118.

Burns, A., Frank, T., Hamill, I. & Shi, J.-M., 2004. The Favre averaged drag model for turbulent dispersion in Eulerian multi-phase flows. *Proceedings of the 5th International Conference on Multiphase Flow.*

Clift, R., Grace, J. & Weber, M., 1978. Bubbles, Drops, and Particles. *Dover Publications.*

Colombo, M. & Fairweather, M., 2015. Multiphase turbulence in bubbly flows: RANS simulations. *International Journal of Multiphase Flow,* Volume 77, pp. 222-243.

Crowe, C., 2000. On models for turbulence modulation in fluid-particle flows. *International Journal of Multiphase Flow,* Volume 26, pp. 719-727.

Crowe, C. T., Troutt, R. & Chung, J. N., 1996. Numerical models for two-phase turbulent flows. *Annual Review of Fluid Mechanics,* Volume 28, pp. 11-43.

Dabiri, S., Lu, J. & Tryggvason, G., 2013. Transition between regimes of a vertical channel bubbly upflow due to bubble deformability. *Physics of Fluids,* Volume 25, 102110.

Davidson, L., 1997. *Large eddy simulations: a note on derivation of the equations for the subgrid turbulent kinetic energies,* Chalmers University of Technology, Gothenburg, Sweden: Technical Report No. 97/12, 980904.

de Bertodano, M. L., Lahey, R. T. & Jones, O. C., 1994. Development of a k-epsilon Model for Bubbly Two-Phase Flow. *Journal of Fluids Engineering,* Volume 116, pp. 128-134.

Deen, N. G., Solberg, T. & Hjertager, B., 2001. Large eddy simulation of the Gas–Liquid flow in a square cross-sectioned bubble column. *Chemical Engineering Science,* 56(21-22), pp. 6341-6349.

Deen, N., Hjertager, B. H. & Solberg, T., 2000. *Comparison of PIV and LDA measurement methods applied to the gasliquid flow in a bubble column.* Lisbon, Proceedings of the 10th International Symposium on Applications of Laser Techniques to Fluid Mechanics.

Deen, N., Solberg, T. & Hjertager, B., 2001. Large eddy simulation of the Gas-Liquid flow in a square cross-sectioned bubble column. *Chemical Engineering Science,* Volume 56, pp. 6341-6349.

Dhotre, M., Deen, N., Niceno, B. K. Z. & Joshi, J., 2013. Large Eddy Simulation for Dispersed Bubbly Flows: A Review. *International Journal of Chemical Engineering,* Volume 2013, 343276.

Dhotre, M., Niceno, B. & Smith, B. L., 2008. Large eddy simulation of a bubble column using dynamic sub-grid scale model. *Chemical Engineering Journal,* Volume 136, pp. 337-348.

Bibliography

Drew, D. & Lahey Jr, R., 1981. Phase distribution mechanisms in turbulent two-phase flow in channels of arbitrary cross section. *Journal of Fluid Engineering,* Volume 103, pp. 583-589.

Drew, D. & Passman, S., 1998. Theory of Multicomponent Fluids. *Springer.*

Egorov, Y. & Menter, F., 2008. Development and application of SST–SAS turbulence model in the DESIDER project. In: P. S-H & H. W, eds. *Advances in hybrid RANS–LES modelling. Notes on numerical fluid mechanics and multidisciplinary design, vol. 97.* Berlin: Springer, pp. 261-270.

Erdogan, S. & Wörner, M., 2014. *Analysis and modelling of liquid phase turbulence kinetic energy equation in bubbly flows via direct numerical simulation.* Hamburg, In 2th International Symposium on Multiscale Multiphase Process Engineering.

Ern, P., Risso, F., Fabre, D. & Magnaudet, J., 2012. Wake-induced oscillatory paths of bodies freely rising or falling in fluids. *Annual Review of Fluid Mechanics,* Volume 44, pp. 97-121.

Ervin, E. A. & Tryggvason, G., 1997. The Rise of Bubbles in a Vertical Shear Flow. *J. Fluids Eng.,* Volume 119, pp. 443-449.

Fox, R. O., 2012. Large-Eddy-Simulation Tools for Multiphase Flows. *Annual Review of Fluid Mechanics,* Volume 44, pp. 47-76.

Fox, R. O., 2014. On multiphase turbulence models for collisional fluid-particle flows. *Journal of Fluid Mechanics,* Volume 742, pp. 368-424.

Fröhlich, J., 2006. Large Eddy Simulation turbulenter Strömungen. *Teubner-Verlag.*

Fröhlich, J., Rodi, W., Dewan, A. & Fontes, J., 2002. Large Eddy Simulation of flow around the free end of a circular cylinder. *Numerical Flow Simulation III, Notes on Numerical Fluid Mechanics, 82, Springer,* pp. 191-202.

Fröhlich, J. & von Terzi, D., 2008. Hybrid LES/RANS Methods for the Simulation of Turbulent Flows. *Progress in Aerospace Sciences,* Volume 44, pp. 349-377.

Germano, M., Piomelli, U., Moin, P. & Cabot, W., 1991. A dynamic subgrid–scale eddy viscosity model. *Physics of Fluids A,* Volume 3, pp. 1760-1765.

Geurts, B. & Fröhlich, J., 2002. A framework for predicting accuracy limitations in large-eddy simulation. *Physics of Fluids,* Volume 14, L41.

Gore, R. & Crowe, C., 1989. Effect of particle size on modulating turbulent intensity. *International Journal of Multiphase Flow,* Volume 15, pp. 279-285.

Göz, M., Bunner, B., Sommerfeld, M. & Tryggvason, G., 2001. *Direct numerical simulation of bidisperse bubble swarms.* New Orleans, International Conference on Multiphase Flow.

Göz, M. & Sommerfeld, M., 2004. Analysis of bubble interactions in bidisperse bubble swarms by direct numerical simulation. *Bubbly Flows. Heat and Mass Transfer, Springer,* pp. 175-190.

Hanjalić, K. & Launder, B., 2011. Modelling Turbulence in Engineering and the Environment. *Cambridge University Press.*

Hosokawa, S. & Tomiyama, A., 2004. Turbulence modification in gas-liquid and solid- liquid dispersed two-phase pipe flows. *International Journal of Heat and Fluid Flow,* Volume 25, pp. 489-498.

Hosokawa, S. & Tomiyama, A., 2010. *Effects of bubbles on turbulent flows in vertical channels.*. Tampa, USA, in 7th International Conference on Multiphase Flow.

Hosokawa, S., Tomiyama, A., Misaki, S. & Hamada, T., 2002. Lateral Migration of Single Bubbles Due to the Presence of Wall. Volume ASME Conf. Proc. 2002 Vol. 1: Fora, Parts A and B / Advances in Free Surface and interface Fluid Dynamics, 855.

Ilić, M., 2006. Statistical Analysis of Liquid Phase Turbulence based on Direct Numerical Simulations of Bubbly Flows. *PhD thesis, Forschungzentrum Karlsruhe.*

Ishii, M. & Hibiki, T., 2011. *THERMO-FLUID DYNAMICS OF TWO-PHASE FLOW.* 2nd ed. New York: Springer Science+Business Media, Inc..

Ishii, M. & Zuber, N., 1979. Drag Coefficient and Relative Velocity in Bubbly, Droplet or Particulate Flows. *AlChE Journal,* Volume 25, pp. 843-855.

Islam, A. A., Adoo, N. & Bergstrom, D., 2016. Prediction of mono-disperse gas–liquid turbulent flow in a vertical pipe. *International Journal of Multiphase Flow,* Volume 85, pp. 236-244.

Jakirlić, S. & Hanjalić, K., 2002. A new approach to modelling near-wall turbulence energy and stress dissipation. *Journal of Fluid Mechanics,* Volume 459, pp. 139-166.

Jakirlić, S. & Jovanović, J., 2010. On unified boundary conditions for improved predictions of near-wall turbulence. *Journal of Fluid Mechanics,* Volume 656, pp. 530-539.

Juliá, J. E., Hernandez, L., Chiva, S. & Vela, A., 2007. Hydrodynamic characterization of a needle sparger rectangular bubble column: Homogeneous flow, static bubble plume and oscillating bubble plume. *Chemical Engineering Science ,* 62(22), pp. 6361-6377.

Kataoka, I. & Serizawa, A., 1989. Basic equations of turbulence in gas-liquid two-phase flow. *International Journal of Multiphase Flow,* Volume 15, pp. 843-855.

Kempe, T. & Fröhlich, J., 2012. An improved immersed boundary method with direct forcing for the simulation of particle laden flow. *Journal of Computational Physics,* Volume 231, pp. 3663-3684.

Kim, J., Moin, P. & Moser, R., 1987. Turbulence statistics in fully developed channel flow at low Reynolds number. *Journal of Fluid Mechanics,* Volume 177, pp. 133-166.

Krepper, E. et al., 2008. The inhomogeneous MUSIG model for the simulation of polydispersed flows. *Nuclear Engineering and Design,* Volume 238, pp. 1690-1702.

Krepper, E., Lucas, D. & Prasser, H.-M., 2005. On the modelling of bubbly flow in vertical pipes. *Nuclear Engineering and Design,* Volume 235, pp. 597-611.

Krepper, E. et al., 2007. Experimental and numerical studies of void fraction distribution in rectangular bubble columns. *Nuclear Engineering and Design,* Volume 237, pp. 399-408.

Lance, M. & Bataille, J., 1991. Turbulence in the liquid phase of a uniform bubbly air-water flow. *Journal of Fluid Mechanics,* Volume 222, pp. 95-118.

Liao, Y., Rzehak, R., Lucas, D. & Krepper, E., 2015. Baseline closure model for dispersed bubbly flow: bubble coalescence and breakup. *Chemical Engineering Science,* Volume 122, pp. 336-349.

Lilly, D., 1992. A proposed modification of the Germano subgrid-scale closure method. *Physics of Fluids A,* Volume 4, pp. 633-635.

Lucas, D. & Tomiyama, A., 2011. On the role of the lateral lift force in poly-dispersed bubbly flows. *International Journal of Multiphase Flow,* Volume 37, pp. 1178-1190.

Lu, J. & Tryggvason, G., 2013. Dynamics of nearly spherical bubbles in a turbulent channel upflow. *Journal of Fluid Mechanics,* Volume 732, pp. 166-189.

Ma, M., Lu, J. & Tryggvason, G., 2015. Using statistical learning to close two-fluid multiphase flow equations for a simple bubbly system. *Physics of Fluids,* Volume 27, 092101.

Mansour, N. N., Kim, J. & Moin, P., 1988. Near-wall k-ε turbulence modelling. *AIAA Journal,* 27(8), pp. 1068-1073.

Martinez Mercado, J. et al., 2010. On bubble clustering and energy spectra in pseudo-turbulence. *Journal of Fluid Mechanics,* Volume 650, pp. 287-306.

Ma, T., Lucas, D., Jakirlić, S. & Fröhlich, J., 2017b. A second-moment closure for bubble-induced turbulence based on DNS data. *(in preparation).*

Ma, T. et al., 2015b. Scale-Adaptive Simulation of a square cross-sectional bubble column. *Chemical Engineering Science,* Volume 131, pp. 101-108.

Ma, T. et al., 2017. Direct numerical simulation–based Reynolds-averaged closure for bubble-induced turbulence. *Physical Review Fluids,* Volume 2, 034301.

Ma, T., Ziegenhein, T., Lucas, D. & Fröhlich, J., 2016. Large eddy simulations of the gas-liquid flow in a rectangular bubble column. *Nuclear Engineering and Design,* Volume 299, pp. 146-153.

Ma, T. et al., 2015a. Euler-Euler large eddy simulations for dispersed turbulent bubbly flows. *International Journal of Heat and Fluid Flow,* Volume 56, pp. 51-59.

Mazzitelli, I. M. & Lohse, D., 2009. Evolution of energy in flow driven by rising bubbles. *Physical Review E,* Volume 79, 066317.

Mendez-Diaz, S., Serrano-Garcia, J. C., Zenit, R. & Hernandez-Cordero, J. A., 2013. Power spectral distributions of pseudo-turbulent bubbly flows. *Physics of Fluids,* Volume 25, 043303.

Meneveau, C., 1994. Statistics of turbulence subgrid-scale stresses: necessary conditions and experimental tests. *Physics of Fluids,* Volume 6, pp. 815-833.

Menter, 2013. private communication.

Menter, 2016. private communication.

Menter, F., 2012. Best Practice: Scale-Resolving Simulations in ANSYS CFD, Version 1.03.

Menter, F. & Egorov, Y., 2005. A Scale-Adaptive simulation model using two-equation models. *AIAA Paper,* Volume 2005-1095.

Menter, F. & Egorov, Y., 2010. Scale-Adaptive simulation Method for Unsteady Flow Predictions. Part 1: Theory and Model Description. *Flow Turbulence and Combustion,* Volume 85, pp. 113-138.

Menter, F. R., 1994. Two-equation Eddy-viscosity turbulence models for engineering applications. *AIAA Journal,* Volume 32, pp. 1598-1605.

Mercado, J. et al., 2010. On bubble clustering and energy spectra in pseudo-turbulence. *Journal of Fluid Mechanics,* Volume 650, pp. 287-306.

Morel, C., 1997. *Turbulence modeling and first numerical simulations in turbulent two-phase flows,* Grenoble, France: SMTH/LDMS/97-023.

Mudde, R. F., 2005. Gravity-driven Bubbly Flows. *Annual Review of Fluid Mechanics,* 37(1), pp. 393-423.

Mudde, R. F., Lee, D. J., Reese, J. & Fan, L.-S., 1997. Role of Coherent Structures on Reynolds Stresses in a 2-D Bubble Column. *Particle Technology and Fluidization,* 43(4), pp. 913-926.

Niceno, B., Boucker, M. & Smith, B. L., 2009. Euler-Euler large eddy simulation of a square cross-sectional bubble column using the Neptune CFD code. *Science and Technology of Nuclear Installations,* Volume 2009.

Niceno, B., Dhotre, M. & Deen, N., 2008. One-equation sub-grid scale (SGS) modelling for Euler--Euler large eddy simulation (EELES) of dispersed bubbly flow. Volume 63, pp. 3923-3931.

Pfleger, D. & Becker, S., 2001. Modelling and simulation of the dynamic flow behaviour in a bubble column. *Chemical Engineering Science,* Volume 56, pp. 1737-1747.

Pfleger, D., Gomes, S., Gilbert, N. & Wagner, H.-G., 1999. Hydrodynamic simulations of laboratory scale bubble columns fundamental studies of the Eulerian-Eulerian modelling approach. *Chemical Engineering Science,* 54(21), pp. 5091-5099.

Politano, M., Carrica, P. & Converti, J., 2003. A model for turbulent polydisperse two-phase flow in vertical channels. *International Journal of Multiphase Flow,* Volume 29, pp. 1153-1182.

Pope, S. B., 2000. *Turbulent Flows.* 1st ed. Cambridge: Cambridge University Press.

Prakash, V. N. et al., 2016. Energy spectra in turbulent bubbly flows. *Journal of Fluid Mechanics,* Volume 791, pp. 174-190.

Prosperetti, A., 2015. Private communication.

Prosperetti, A. & Tryggvason, G., 2007. Computational Methods for Multiphase Flow. *Cambridge University Press.*

Rensen, J., Luther, S. & Lohse, D., 2005. The effect of bubbles on developed turbulence. *Journal of Fluid Mechanics,* Volume 538, pp. 153-187.

Riboux, G., Legendre, D. & Risso, F., 2013. A model of bubble-induced turbulence based on large-scale wake interactions. *Journal of Fluid Mechanics,* Volume 719, pp. 362-387.

Riboux, G., Risso, F. & Legendre, D., 2010. Experimental characterization of the agitation generated by bubbles rising at high Reynolds number. *Journal of Fluid Mechanics,* Volume 643, pp. 509-539.

Risso, F. & Ellingsen, K., 2002. Velocity fluctuations in a homogeneous dilute dispersion of high-Reynolds-number rising bubbles. *Journal of Fluid Mechanics,* Volume 453, pp. 395-410.

Rodi, W. & Mansour, N. N., 1993. Low Reynolds number k-epsilon modelling with the aid of direct simulation data. *Journal of Fluid Mechanics,* Volume 250, pp. 509-529.

Roghair, I. et al., 2011b. On the drag force of bubbles in bubble swarms at intermediate and high Reynolds numbers. *Chemical Engineering Science,* Volume 66, pp. 3204-3211.

Roghair, I., M.W.Baltussen, Annaland, M. S. & Kuipers, J., 2013. Direct Numerical Simulations of the drag force of bi-disperse bubble swarms. *Chemical Engineering Science,* Volume 95, pp. 48-53.

Roghair, I. et al., 2011a. Energy spectra and bubble velocity distributions in pseudo-turbulence: Numerical simulations vs. experiments. *International Journal of Multiphase Flow,* Volume 37, pp. 1093-1098.

Rzehak, R. & Krepper, E., 2013. CFD modeling of bubble-induced turbulence. *International Journal of Multiphase Flow,* Volume 55, pp. 138-155.

Santarelli, C., 2015. *Direct Numerical Simulations of Bubbles in Turbulent Flows with Heat Transfer (Ph.D.thesis).* Technische Universität Dresden: TUD press.

Santarelli, C. & Fröhlich, J., 2015a. Direct Numerical Simulations of spherical bubbles in vertical turbulent channel flow. *International Journal of Multiphase Flow,* Volume 75, pp. 174-193.

Santarelli, C. & Fröhlich, J., 2015b. Private communication.

Santarelli, C. & Fröhlich, J., 2016b. Direct Numerical Simulations of spherical bubbles in vertical turbulent channel flow. Influence of bubble size and bidispersity. *International Journal of Multiphase Flow,* Volume 81, pp. 27-45.

Santarelli, C., Roussel, J. & Fröhlich, J., 2016a. Budget analysis of the turbulent kinetic energy for bubbly flow in a vertical channel. *Chemical Engineering Science,* Volume 141, pp. 46-62.

Sato, Y., Sadatomi, M. & Sekoguchi, K., 1981. Momentum and heat transfer in two-phase bubble flow I: Theory. *International Journal of Multiphase Flow,* Volume 7, pp. 167-177.

Sato, Y. & Sekoguchi, K., 1975. Liquid velocity distribution in two-phase bubble flow. *International Journal of Multiphase Flow,* Volume 2, pp. 79-95.

Schiller, L. & Naumann, A., 1933. Über die grundlegenden Berechungen bei der Schwerkraftaufbereitung. *Ver. Deut. Ing.,* Band 77, pp. 318-320.

Serizawa, A., Kataoka, I. & Michiyoshi, I., 1975. Turbulence structure of air-water bubbly flow - II. local properties. *International Journal of Multiphase Flow,* Volume 2, pp. 235-246.

Shur, M., Spalart, P., Strelets, M. & Travin, A., 2008. A hybrid RANS-LES approach with delayed-DES and wall-modeled LES capabilities. *International Journal of Heat and Fluid Flow,* Volume 29, pp. 1638-1649.

Smagorinsky, J., 1963. General circulation experiments with the primitive equations. I. The basic experiment.. *Monthly Weather Review,* Volume 91, pp. 99-164.

Sungkorn, R., Derksen, J. J. & Khinast, J. G., 2011. Modeling of turbulent gas-liquid bubbly flows using stochastic Lagrangian model and lattice-Boltzmann scheme. *Chemical Engineering Science,* Volume 66, pp. 2745-2757.

Tabib, M., Roy, S. & Joshi, J., 2008. CFD simulation of bubble column—An analysis of interphase forces and turbulence models. *Chemical Engineering Journal,* Volume 139, pp. 589-614.

Tanaka, T. & Eaton, J. K., 2008. Classification of Turbulence Modification by Dispersed Spheres Using a Novel Dimensionless Number. *Physical Review Letters,* Volume 101, 114502.

Tanaka, T. & Eaton, J. K., 2010. Sub-Kolmogorov resolution PIV measurements of turbulence modification by particles. *Journal of Fluid Mechanics,* Volume 643, pp. 177-206.

Tomiyama, A., 2002. *Single Bubbles in Stagnant Liquids and in Linear Shear Flows.* Dresden, Germany, Workshop on Measurement Technology (MTWS5).

Tomiyama, A., Kataoka, I., Zun, I. & Sakaguchi, T., 1998. Drag coefficients of single bubbles under normal and micro gravity conditions. *JSME International Journal Series B,* 41(2), p. 472.

Tomiyama, A. et al., 1995. Effects of Eötvös number and dimensionless liquid volumetric flux on lateral motion of a bubble in a laminar duct flow. In: A. Serizawa, T. Fukano & J. Bataille, eds. *Advances in multiphase flow.* : Elsevier, pp. 3-15.

Tomiyama, A., Tamai, H., Zun, I. & Hosokawa, S., 2002. Transverse migration of single bubbles in simple shear flows. *Chemical Engineering Science ,* 57(11), pp. 1849-1858.

Troshko, A. A. & Hassan, Y. A., 2001. A two-equation turbulence model of turbulent bubbly flows. *International Journal of Multiphase Flow,* Volume 27, pp. 1965-2000.

Tryggvason, G., Scardovelli, R. & Zaleski, S., 2011. Direct Numerical Simulations of Gas-Liquid Multiphase Flow. *Cambridge University Press.*

Uhlmann, M., 2005. An immersed boundary method with direct forcing for the simulation of particulate flows. *Journal of Computational Physics,* Volume 209, pp. 448-476.

Vanga, B. et al., 2005. *Two-fluid model LES of a bubble column.* Avignon, NURETH-11.

Wang, Q. & Yao, W., 2016. Computation and validation of the interphase force models for bubbly flow. *International Journal of Heat and Mass Transfer ,* Volume 98, pp. 799-813.

Welch, P., 1967. The Use of Fast Fourier Transform for the Estimation of Power Spectra: A Method Based on Time Averaging Over Short, Modified Periodograms. *IEEE Transactions on Audio Electroacoustics,* Volume AU-15, pp. 70-73.

Wellek, R. M., Agrawal, A. K. & Skelland, A. H. P., 1966. Shape of Liquid Drops Moving in Liquid Media. *AlChE Journal,* Volume 12, pp. 854-862.

Wilcox, D. C., 1998. *Turbulence Modeling for CFD.* 2nd ed. La Canada, California: DCW Industries, Inc..

Wörner, M. & Erdogan, S., 2013. Toward improved closure relations for the turbulent kinetic energy equation in bubble-driven flows. *Chemie Ingenieur Technik,* Volume 85 (7), pp. 1131-1136.

Yang, Z. & Shih, T. H., 1993. New Time Scale Based k-ε Model for Near-Wall Turbulence. *AIAA Journal,* Volume 31, pp. 1191-1198.

Zhang, D., Deen, N. & Kuipers, J., 2006. Numerical simulation of the dynamic flow behavior in a bubble column: a study of closures for turbulence and interface forces. *Chemical Engineering Science,* Volume 61, pp. 7593-7608.

Ziegenhein, T. & Lucas, D., 2016. On sampling bias in multiphase flows: Particle image velocimetry in bubbly flows. *Flow Measurement and Instrumentation,* Volume 48, pp. 36-41.

Ziegenhein, T., Rzehak, R. & Lucas, D., 2015. Transient simulation for large scale flow in bubble columns. *Chemical Engineering Science,* Volume 122, pp. 1-13.